CW01347525

Cooking Technology

Cooking Technology

Transformations in Culinary Practice
in Mexico and Latin America

EDITED BY
STEFFAN IGOR AYORA-DIAZ

Bloomsbury Academic
An imprint of Bloomsbury Publishing Plc

BLOOMSBURY
LONDON · OXFORD · NEW YORK · NEW DELHI · SYDNEY

Bloomsbury Academic
An imprint of Bloomsbury Publishing Plc

50 Bedford Square 1385 Broadway
London   New York
WC1B 3DP   NY 10018
UK   USA

www.bloomsbury.com

Bloomsbury and the Diana logo are trademarks of Bloomsbury Publishing Plc

First published 2016

© Selection and Editorial Material: Steffan Igor Ayora-Diaz, 2016
© Individual Chapters: Their Authors, 2016

Steffan Igor Ayora-Diaz has asserted his right under the Copyright, Designs and Patents Act, 1988, to be identified as Author of this work.

All rights reserved. No part of this publication may be reproduced or transmitted in any form or by any means, electronic or mechanical, including photocopying, recording, or any information storage or retrieval system, without prior permission in writing from the publishers.

No responsibility for loss caused to any individual or organization acting on or refraining from action as a result of the material in this publication can be accepted by Bloomsbury or the author.

British Library Cataloguing-in-Publication Data
A catalogue record for this book is available from the British Library.

ISBN: HB: 978-1-4742-3468-9
       ePDF: 978-1-4742-3470-2
       ePub: 978-1-4742-3469-6

Library of Congress Cataloging-in-Publication Data
A catalog record for this book is available from the Library of Congress.

Typeset by RefineCatch Ltd, Bungay, Suffolk, UK
Printed and bound in Great Britain

Contents

Acknowledgments viii

Introduction—The meanings of cooking and the kitchen: Negotiating techniques and technologies 1
Steffan Igor Ayora-Diaz

PART ONE Refiguring the past, rethinking the present 13

1. Grinding and cooking: An approach to Mayan culinary technology 15
Lilia Fernández-Souza

2. Technology and culinary affectivity among the Ch'orti' Maya of Eastern Guatemala 29
Julián López García and Lorenzo Mariano Juárez

3. From bitter root to flat bread: Technology, food, and culinary transformations of cassava in the Venezuelan Amazon 41
Hortensia Caballero-Arias

4. Technologies and techniques in rural Oaxaca's Zapotec kitchens 55
Claudia Rocío Magaña González

PART TWO   Transnational and translocal meanings 69

5   The Americanization of Mexican food and change in cooking techniques, technologies, and knowledge 71
    Margarita Calleja Pinedo

6   Home kitchens: Techniques, technologies, and the transformation of culinary affectivity in Yucatán 85
    Steffan Igor Ayora-Diaz

7   If you don't use chilies from Oaxaca, is it still *mole negro*? Shifts in traditional practices, techniques, and ingredients among Oaxacan migrants' cuisine 99
    Ramona L. Pérez

8   Changing cooking styles and challenging cooks in Brazilian kitchens 111
    Jane Fajans

9   Global dimensions of domestic practices: Kitchen technologies in Cuba 125
    Anna Cristina Pertierra

PART THREE   Recreating tradition and newness 137

10   Recipes for crossing boundaries: Peruvian fusion 139
     Raúl Matta

11   Forms of Colombian cuisine: Interpretation of traditional culinary knowledge in three cultural settings 153
     Juliana Duque Mahecha

12 Cooking techniques as markers of identity and authenticity in Costa Rica's Afro-Caribbean foodways 167
Mona Nikolić

Afterword 181
Carole Counihan

Notes on contributors 185
Index 189

Acknowledgments

This book follows on from a session I organized at the 2013 annual meeting of the American Anthropological Association. Unfortunately, not all who presented papers at that session were able to participate in this volume, but I would nevertheless like to thank Edward Fischer for his contribution on the day and to all those who have joined the conversation subsequently. I am also grateful to Molly Beck, Abbie Sharman, and Jennifer Schmidt at Bloomsbury, and to Lisa Carden, for their interest and support in taking this project through to publication, as well as to the anonymous reviewers who commented on it.

This book has been made possible thanks to the support of the former Chair of the Facultad de Ciencias Antropológicas at the Universidad Autónoma de Yucatán, Dr. Genny Mercedes Negroe Sierra, who allowed me to focus on the project. I have been able to complete this work as a result of the economic support of PIFI (Programa Integral de Fortalecimiento Institucional, Integral Program for Institutional Reinforcement, from the Mexican Secretariat of Education), which allowed me a brief residence as Research Associate at the University of Indiana, Bloomington. During my stay I enjoyed the hospitality of Catherine Tucker, Chair of the Department of Anthropology, of Dan Knudsen, Chair of the Department of Geography, and of Richard Wilk and Anne Pyburn in the Department of Anthropology.

Steffan Igor Ayora-Diaz
Universidad Autónoma de Yucatán

Introduction

The meanings of cooking and the kitchen:

Negotiating techniques and technologies

Steffan Igor Ayora-Diaz, Universidad Autónoma de Yucatán

Various studies have demonstrated the importance of food and its multiple meanings for establishing social and communal ties, as well as its use as a tool to mark the boundaries between groups of people (e.g., Ayora-Diaz 2012; Counihan 2004; Counihan and Kaplan 1998; Heatherington 2001; Holtzman 2009; Montanari 2007). Our main objective in this volume is to examine the types of changes that occur within the space of the kitchen and that often lead to modifications in cooking practices, which then translate into changes in the taste and meanings of food. The following chapters seek to contribute to a more nuanced understanding of contemporary kitchens and cooking in Mexico and Latin America. Latin America in general, and its kitchens in particular, have frequently been represented as technologically backward sites; as places where tradition sits uncontested. More specifically, the dominant focus of attention placed on rural, peasant, and ethnic groups has contributed to perpetuate this image, neglecting the fact that even the most isolated groups are undergoing constant change as a result of their direct or indirect insertion in broader social, economic, political, and cultural processes. Several chapters in this volume seek to challenge this misrepresentation and partial understanding, and show through anthropological, archaeological, and ethno-historical lenses that Latin American kitchens have been and are places where the meanings of food, techniques, and technologies, as well as associated aesthetic values, are endlessly negotiated. The book critically examines the places and times when "traditional" and "modern" culinary values are maintained or negotiated, from the valorization of traditional ingredients, techniques, and technologies—and the nostalgia they sometimes either trigger or to which they respond—to the acceptance and promotion of fusion foods in different urban environments where consumerist practices influence food re-creation. Here we show that in Latin American kitchens, "tradition" and "modernity" are continuously re-signified.

Contributors to this volume are either sociocultural anthropologists or scholars working closely with anthropological texts and issues. To date, there has been little appetite for the study of the kitchen and of cooking: indeed, when some anthropologists look at kitchens, they see a marginal space that plays host to everyday drudgery, and where the actions performed within are restricted to creating something edible from outdoor human action (running the whole gamut from hunting, gathering, horticulture, and agriculture to supermarket shopping). Anthropologists have tended to look at the symbolic significance of the food prepared within that space (again, in the public space), the meaning of the kitchen as an intimate space of the house for family members and their close friends, or to the symbolic meaning of ceremonies and rituals where special, extraordinary meals, are consumed. For example, Mary Douglas (1975) paid attention to the symbolic organization of family meals; Claude Lévi-Strauss (1968) argued for the importance of the culinary "triangle"; Arjun Appadurai (1981) for the political meaning of food in domestic and ritual spaces; and Paul Stoller (1989) described how women could use the taste of food in domestic/inter-ethnic political contexts. Yet, in these examples there is very little attention paid to the technology and the organization of practices required to produce everyday meals.

Examples from Latin America reflect the same bias. Traditional anthropological accounts have privileged the study of rural, peasant, and impoverished urban people. Hence the focus has been, for example, on the social organization of the lowland Maya of the Yucatán peninsula and of villages in the mountains of Chiapas (Redfield 1940; Nash 1970; even when they describe the house, they only list the cooking implements but do not describe how they are used. See, for example, Redfield 1946: 12–16); on the domestic economy of guinea pig (*cuy*) consumption in the Andes, where although recognizing its economic and symbolic significance, the author pays little attention to its preparation (e.g., Morales 1995); on the gendered organization of cooking during the preparation of ceremonial foods in central Mexico (Christie 2011); and on the economic role of women in Peruvian markets (Babb 1989). These publications examine the "traditional" structure of villages or the effects of capitalism and gender politics on the social production and consumption of food. Without denying their importance, they again pay scant attention to the techniques and technologies required for the preparation of food, and to the processes that have made available new ingredients that vie to displace "traditional" ones. Other studies privilege the description of antiquated technologies to augment the contrast with the "modern" world and its kitchens, disregarding the fact that often those kitchens, the appliances they use, and their ingredients have either withstood globally-driven processes or changed as a result of them. It is only in the last two decades that anthropologists have started to focus specifically on the preparation and consumption of food as part of larger social, economic, and political processes, as the chapters in this collection do. The authors represented in this volume share the conviction that we also need to examine how technological and technical changes are gradually or rapidly introduced in different contexts and thus challenge the received wisdom that cooks tend to be conservative in Latin American contexts.

The kitchen is an anthropological problem

As is evident through the different chapters of this book, we do not understand the kitchen as an isolated space where practices are shaped independently of the world beyond its walls (or roof, in some cases). On the contrary, we take the kitchen as a privileged space where global, local, and translocal transformations in the circulation of edible and culinary technologies converge and refashion each other through everyday culinary techniques. These transformations contribute to change, in turn, the meanings of the space of the kitchen itself, and its importance within the home, the place, and region, and of the ingredients and technologies that cooks use in them. David Harvey (1990), Roland Robertson (1992) and Arjun Appadurai (1996), among others, have used different concepts to describe and analyze the types and modes in which these supplementary global transformations interact in the context of contemporary globalization.

There is already a growing preoccupation with social, class, and gender politics, as well as with the political economic transformations that foster changes in the kitchens of North Atlantic societies, from the introduction of new appliances to the reorganization of cooking spaces (Shove et al. 2008; Freeman 2004). Within anthropology, the focus on material culture has allowed for an examination of the part played by different domestic appliances in contemporary everyday life, including those that are usually enclosed within kitchen walls (Miller 2002; Pink 2004). Regarding these appliances, there is an ongoing debate about their emancipating effects not only for women, but for all family members (Cowan 1983; Rutherford 2003). In 1948, Sigfried Giedion

(1948) suggested that the transposition of mechanization and its modes of organization from the factories to the home were to have liberating effects for all members of the family. However, most studies seem to suggest that although appliance acquisition and use may have led to many benefits and to some degree of reorganization of domestic work (shaped as well by the increasing need for two-income family budgets), in general they have failed to deliver freedom in or from the kitchen (Rees 2013; Silva 2010). Their effects on the transformation of gender roles are contentious as well (Chabaud-Rychter 1994; Cockburn and Fürst-Dilic 1994; Ormond 1994). While most authors who examine domestic and kitchen transformations do so in North Atlantic societies, it is necessary to acknowledge that Latin American and Caribbean cultures have also been heavily involved in those processes we have named "modernization" and "globalization." In this sense, this volume seeks to address this imbalance and to spark further questions for research focusing on the kitchen and the house as privileged settings where complex processes intersect. Despite its seemingly profoundly local nature, the kitchen features appliances, tools, instruments, and ingredients that follow global-local paths before they finally enter the "enclosed" space of the home. These processes mobilize cooking instruments, tools, electric and electronic technologies, and appliances, some of which are included in everyday culinary practices, while others are adapted or rejected. As a result of the everyday interaction among these technologies, and that between cooks and their technologies, changes in culinary practices and cooking techniques sometimes emerge, leading to a range of effects on the aesthetics of food and, therefore, on everyday life in general (Ayora-Diaz 2014).

The problem of technology in the anthropology of food

In anthropology there are many definitions of "technology," and the chapters in this volume reflect that diversity. For example, separating technology from culture, Allen W. Batteau suggests that technology can be given a restrictive meaning excluding tools and instruments, arguing that an encompassing definition is useless (2010: 3). In contrast, proposing a broader definition, Nathaniel Schlanger (2006) questions the common anthropological distinction between techniques as the stuff of "simple" societies, while technologies are found in "modern" societies. He suggests that, in accordance with its original meaning (the study of techniques), we need to embrace the study of techniques and practices as part of our disciplinary understanding of technology. For his part, Mike Michael (2006) underscores the tight relationship between science and technology calling *technoscientific* those everyday objects that result from the direct application of scientific knowledge. Studying the kitchen demands we broaden our understanding of "technology" to encompass different objects, instruments, devices, appliances, techniques, and other electric and electronic technologies. Everyday practices in the kitchen are often structured by implicit and explicit rules derived from written instruction in manuals of home economics and cookbooks, in addition to those received from oral tradition, highlighting the connections between broader social processes and the extremely localized practices of the kitchen (Bower 1997; Goldstein 2012; Rutherford 2003). In the same way in which a microwave oven, an electric steamer, or the refrigerator can lead us to transform our cooking practices, so does the widespread availability of processed, mass-produced, prepackaged ingredients and meals. Taken together, these items also change our taste and appreciation for everyday foods, as well as their meaning for the society that originally produced and consumed them. Consequently, we need a concept of technology that is broad enough to help us:

- understand cookbooks as technologies of inscription that support our culinary memory (including, increasingly, the use of electronic devices to access recipes online);
- recognize instruments, appliances, and tools as culturally appropriate or inappropriate technologies for food preparation, in each localized society and each kitchen;
- also understand the mediating role played by ingredients as technological objects that engender the relationship between science, technology, ingredients, and cultural norms through the preparation of everyday meals. For example, the use of either industrially processed or organic foodstuffs in the preparation of dinner, or the consumption of prepackaged foods, shows our relations with and our ethics about different technologies. Again, the use of both processed and organic ingredients can be seen as the product of local–global and translocal interactions mediated by commercial, political, and ethical principles that, in turn, mediate their forms of localized adoption.

This volume discusses culinary transformations in relation to the arrival of new appliances and technologies, the availability of different ingredients and processed meals, the spread of *high* culinary values through restaurants that specialize in hautes or nouveaux cuisines, the pervasive broadcast of TV gourmet programs, and the growing access to cookbooks specialized in a broad range of national, regional, and so-called "international" cookery styles. Consequently, we include tools, instruments, cookbooks and cooks' magazines, ingredients, electric appliances, and electronic media, which lead, through their interaction, to different modes of culinary transformation. Given that contemporary global society, in both urban and rural areas, is marked by continuous change, we find that in some cases these technological appropriations and developments may lead to the radical transformation of the food of a society. In other cases they lead to a nostalgic understanding of the past, which is expressed in the reclamation of traditional techniques, technologies, and tastes. In most cases, however, whether transforming their food into radically new forms, or through attempts to recover and affirm a culinary or gastronomic tradition, subjects must engage in complex negotiations.

As contributors demonstrate throughout the book, different actors may intervene in these processes, from state institutions to food and agribusiness corporations, printed, electronic, online and televised media, cookbook writers, and celebrity chefs who appear on TV, radio and newspapers. Moreover, restaurants, the manufacturers and vendors of domestic appliances, cookery schools, seminars and workshops, and even following the lead of urban eateries and street food vendors, are turned into catalysts for change in the kitchen and the home. Each has a mediating part to play in the different cases presented here, so that even communities portrayed as "isolated" are included in global processes in which groups of tourists, driven by nostalgia and searching for authenticity, may demand a "return" to old techniques and taste.

The structure of this book

This book is divided into three sections within which the chapters explore, from different approaches, the diversity of meanings that the kitchen and the technologies and culinary practices performed therein yield over time. Each chapter focuses on specific problems in different parts of Mexico, the United States, and Latin America, and although some themes are present in every section, the chapters are ordered by their primary preoccupation. Part One, "Refiguring the past, rethinking the present," includes four chapters in which the relationship with the past, real and imagined, is the strongest component of the argument. Thus, in Chapter 1, Lilia Fernández-Souza examines, from a predominantly ethno-archaeological perspective, the ways in which the study of the space of the kitchen has been approached among the Maya, paying particular attention to the use of milling stones (*metates*) and mortars (*molcajetes*) and the use of underground ovens (*pibs*). She attempts to establish whether there are elements of continuity between the Maya of the past and those of the present. In Chapter 2, Julian López García and Lorenzo Mariano Juárez critically examine the top-down attempts to transform culinary technologies among the Ch'orti' of Guatemala. They focus on the everyday use of milling stones (*metates*) and clay griddles (*comales*), arguing that development agencies have found strong resistance to proposed changes to cooking appliances, not because of some traditionally conservative rejection of new technologies, but because the promoters of development have failed to understand the *meaning* of long-established tools, and the affects they mobilize in everyday life. In Chapter 3, Hortensia Caballero-Arias examines the cassava production process undertaken by Venezuelan Yanomami, demonstrating how this root has become deeply ingrained in the imagination of "primitive" indigenous societies, thus inhibiting its incorporation into urban "modern" diets. However, as she also notes, the recuperation of "traditional" ingredients as a strategy to revalorize national cuisines is gradually changing urban consumers' relationship to this root. Finally, in Chapter 4 Claudia Rocío Magaña González describes the negotiations required to explain the fusion of "modern" and "traditional" technologies in the preparation of Zapotec, Istmeño cuisine in Oaxaca. She shows how even in "traditional" settings, cooks are willing to introduce changes in the use of cooking instruments and appliances while maintaining their affective relationship to "traditional" foods.

The five chapters included in Part Two, "Transnational and translocal meanings," place the emphasis on global-local and translocal connections resulting from the movement of people, of the border between nations, and of culinary commodities, as they are expressed in the space of the kitchen and mediated by the use of technologies and culinary techniques. In Chapter 5, Margarita Calleja Pinedo examines how, in the southern United States, changes over time to the Mexican dish *carne con chile* resulted in one of the mainstays of contemporary Tex-Mex food, *chili con carne*. Her chapter examines the role played by the transition from oral to written recipes and how emerging businesses specialized in the industrial packaging of ingredients and meals. She shows how the understanding of this food changed with the growing appropriation of this originally Mexican recipe by Anglophone society in the US. In Chapter 6, Steffan Igor Ayora-Diaz examines the contemporary transformations of urban life in Yucatán, arguing that converging transformations in the foodscape, the space of kitchens, the diversity of cookbooks, the arrival of new cooking appliances and ingredients have led to changes in culinary techniques, in the taste of Yucatecan food, and in the meaning and local affective attachments to regional food. In Chapter 7, Ramona L. Pérez examines the culinary techniques and technologies employed by migrants from the state of Oaxaca, in southern Mexico, who now live in the south of the US. She discusses the affective relationship of people from Oaxaca with the food of their region and the ways in which they signify and re-signify though nostalgia and "authenticity" foods they consider traditional and meaningful—clearly illustrated by the preparation of *mole negro*. In Chapter 8, Jane Fajans takes us to urban Brazil, a context where middle- and upper-class families have customarily employed maids to cook their meals, but where in recent times the media, celebrity chefs, and other thematic recuperations of "Brazilian" food have encouraged members of well-to-do families to take upon cooking national dishes. In this case, hautes and fusion cuisines are fostering change in how Brazilians adopt national dishes into their everyday culinary and gastronomic practices. To conclude this section, in Chapter 9, Anna Cristina Pertierra examines the transformative process of Cuban kitchens and the appliances that furnish them from the eve of the Cuban revolution to the post-Soviet era, passing through the Soviet control of the island's market. As she argues, the kitchen becomes the locus of negotiation both for political meanings and for the construction of a national identity, making it the ideal space for state intervention and transformation.

The three chapters encompassed in Part Three, "Recreating tradition and newness," examine contemporary transformations derived from the strategic use of nostalgia, tourism development, and haute cuisine in the rediscovery and valorization of ethnic and national cuisines. In these chapters we can see how different technologies are developed, appropriated, or adapted into the space of kitchens to produce results that respond to a global culinary forces. Thus, in Chapter 10 Raúl Matta looks upon the part played by Peruvian celebrity chefs who following contrasting approaches seek to incorporate guinea pigs (*cuy*) into upscale menus. Normally associated with the imagination of indigenous diets and shunned by urbanites, renowned chefs apply new culinary technologies to transform it into a special dish, or modify its presentation seeking to enhance its desirability on the part of their customers. He shows how their efforts gradually induce a greater acceptance for uncommon ingredients in the non-indigenous, urban diet. In Chapter 11, Juliana Duque-Mahecha looks at Colombian foods in three different settings: fine dining restaurants, comfort-food establishments, and food stalls in popular markets. She shows how these three different backdrops converge in the re-signification and revalorization of "traditional" Colombian foods, and contribute to the creation and establishment of a shared image of what constitutes the taste and the components of an "authentic" national cuisine defined and valued as "cultural heritage." Finally, in Chapter 12, Mona Nikolić focuses on an Afro-Caribbean community in Costa Rica. She discusses and illustrates how contemporary changes had led the local population to abandon cooking techniques and technologies, displacing traditional meals from their menus and replacing them with commercially processed ingredients. However, recent efforts to market the village in question as a tourism destination have created a context that favors the re-creation of authenticity, and local actors are today engaged in the recuperation of those dishes and the "proper" techniques and technologies required for their production. The tourist gaze thus plays an important part in the re-signification of "authentic" local meals. The book concludes with an Afterword by Carole Counihan. As she suggests, this volume opens venues for further research:

1 the conceptual and practical convergence of taste, place, technology and access;

2 the flexibility in the adoption and change in culinary practices according to context;

3 the different ways in which technology shapes, and is shaped by, different forms of identity;

4 the continuous negotiation between the adoption and continuation of "traditional" and "modern" technologies. Culinary techniques and technologies clearly require greater attention in future studies.

The chapters collected in this book offer different and complex views on the meanings and the relationships between techniques, technologies, and ingredients that coexist in the space of "traditional" and "modern" kitchens. Both meanings and relationships need to be understood as mediated and negotiated by a diversity of social actors, suggesting that they change in situations of structural inequality and amidst global, global–local, and trans-local processes expressed in the complex production, circulation, and consumption of edible commodities and culinary technologies. Hence, the kitchen in Latin America and Mexico is not a stagnant and closed space, but rather a site where broader and more complex processes converge, transform everyday practices and meanings, and provide a ground for local, regional, ethnic, national, and cosmopolitan affiliations. Our hope is that this volume will trigger further questions in different places of Mexico and Latin America, and bring recognition to the multiple meanings that converge in the kitchen and the meals produced therein.

References

Appadurai, A. (1981), "Gastro-Politics in Hindu South Asia." *American Ethnologist* 8(3):494–511.
Appadurai, A. (1996), *Modernity at Large: Cultural Dimensions of Globalization*. Minneapolis: University of Minnesota Press.
Ayora-Diaz, S. I. (2012) *Foodscapes, Foodfields and Identities in Yucatán*. Amsterdam: CEDLA / New York: Berghahn.
Ayora-Diaz, S. I. (2014), "Estética y tecnología en la cocina meridana." Pp. 317–341 in S. I. Ayora-Diaz and G. Vargas Cetina (eds), *Estética y poder en la ciencia y tecnología: Acercamientos multidisciplinarios*. Mérida: UADY.
Babb, F. (1998 [1989]), *Between Field and Cooking Pot: The Political Economy of Market Women In Peru*. Austin: University of Texas Press.

Batteau, A. W. (2010), *Culture and Technology*. Long Grove: Waveland Press.
Bower, A. L. (ed.) (1997), *Recipes for Reading: Community Cookbooks, Stories, Histories*. Amherst: University of Massachusetts Press.
Chabaud-Rychter, D. (1994), "Women Users in the Design Process of a Food Robot: Innovation in a French Domestic Appliance Company." Pp. 77–93 in C. Cockburn and R. Fürst-Dilic (eds), *Bringing Technology Home. Gender and Technology in a Changing Europe*. Buckingham: Open University Press.
Christie, M. E. (2011), *Kitchenspace: Fiestas and Everyday Life in Central Mexico*. Austin: University of Texas Press.
Cockburn, C. and Fürst-Dilic, R. (1994), "Introduction: Looking for the Gender/Technology Relation." Pp. 1–21 in C. Cockburn and R. Fürst-Dilic (eds), *Bringing Technology Home: Gender and Technology in a Changing Europe*. Buckingham: Open University Press.
Counihan, C. (2004) *Around the Tuscan Table: Food, Family and Gender in Twentieth Century Florence*. New York: Routledge.
Counihan, C. and Kaplan, S. L. (eds) (1998), *Food and Gender*. London: Routledge.
Cowan, R. S. (1983), *More Work for Mother: The Ironies of Household Technology from the Open Hearth to the Microwave*. New York: Basic Books.
Douglas, M. (1997 [1975]), "Deciphering a Meal." Pp. 36–54 in C. Counihan and P. Van Esterick (eds), *Food and Culture: A Reader*. London: Routledge.
Freeman, J. (2004), *The Making of the Modern Kitchen: A Cultural History*. Oxford: Berg.
Giedion, S. (2013 [1948]) *Mechanization Takes Command. A Contribution to Anonymous History*. Minneapolis: University of Minnesota Press.
Goldstein, C. M. (2012) *Creating Consumers: Home Economics in Twentieth-Century America*. Chapel Hill: University of North Carolina Press.
Harvey, D. (1990), *The Condition of Postmodernity. An Enquiry into the Origins of Cultural Change*. Oxford, UK: Blackwell.
Heatherington, T. (2001), "In the Rustic Kitchen: Real Talk and Reciprocity." *Ethnology* 40:309–345.
Holtzman, J. (2009) *Uncertain Tastes: Memory, Ambivalence and the Politics of Eating in Samburu, Northern Kenya*. Berkeley: University of California Press.
Lévi-Strauss, C. (1989 [1968]), "The Culinary Triangle." Pp. 28–35 in C. Counihan and P. Van Esterick (eds), *Food and Culture: A Reader*. London: Routledge.
Michael, M. (2006), *Technoscience and Everyday Life*. London: Open University Press.
Miller, D. (2002), *Home Possessions: Material Culture Behind Closed Doors*. Oxford, UK: Berg.
Montanari, M. (2007) *Il cibo come cultura*. Bari: Laterza.
Morales, E. (1995), *The Guinea Pig: Healing, Food, and Ritual in the Andes*. Tucson: University of Arizona Press.
Nash, J. C. (1970), *In the Eyes of the Ancestors: Belief and Behavior in a Maya Community*. New Haven, CT: Yale University Press.
Ormond, S. (1994) "'Let's Nuke Dinner': Discursive Practices of Gender in the Creation of a New Cooking Process." Pp. 42–58 in C. Cockburn and R. Fürst-Dilic (eds), *Bringing Technology Home: Gender and Technology in a Changing Europe*. Buckingham: Open University Press.

Pink, S. (2004), *Home Truths: Gender, Domestic Objects and Everyday Life.* Oxford: Berg.

Redfield, R. (1940), *The Folk Culture of Yucatán.* Chicago: University of Chicago Press.

Redfield, R. (1946) "Los mayas actuales de la peninsula yucatanense." Pp. 7–30 in C. Torres Novelo and C. A. Echánove Trujillo (eds), *Enciclopedia Yucatanense, Vol VI: Yucatán actual.* Mexico City: Gobierno de Yucatán.

Rees, J. (2013), *Refrigeration Nation: A History of Ice, Appliances, and Enterprise in America.* Baltimore: Johns Hopkins University Press.

Robertson, R. (1992), *Globalization: Social Theory and Global Culture.* London: Sage Publications.

Rutherford, J. W. (2003), *Selling Mrs. Consumer: Christine Frederick and the Rise of Household Efficiency.* Athens, GA: University of Georgia Press.

Schlanger, N. (2006), "Introduction: Technological Commitments: Marcel Mauss and the Study of Techniques in the French Social Sciences." Pp. 1–29 in N. Schlanger (ed.), *Marcel Mauss: Techniques, Technology and Civilization.* New York: Berghahn.

Shove, E., Watson, M., Hand, M., and Ingram, J. (2008), *The Design of Everyday Life.* Oxford, UK: Berg.

Silva, E. B. (2010), *Technology, Culture, Family: Influences on Home Life.* New York: Palgrave MacMillan.

Stoller, P. (1989), *The Taste of Ethnographic Things: The Senses in Anthropology.* Philadelphia: University of Pennsylvania Press.

PART ONE

Refiguring the past, rethinking the present

1

Grinding and cooking:
An approach to Mayan culinary technology

Lilia Fernández-Souza, Universidad Autónoma de Yucatán

Introduction: approaches to Mayan food

Food is one of the most essential parts of daily human life. Thinking about what to eat, how to prepare different sorts of ingredients and dishes, who can eat what kind of food and on what occasions is it to be served are just some of the many choices that a social group, a family, or an individual must make every day. A number of scholars have discussed the role that food and culinary practices play with reference to identity, sociability, prestige, and sense of community (Ayora-Diaz 2012; Juárez López 2008: 19; Scholliers and Claflin 2012: 1), including how culinary traditions stir feelings of nostalgia and nationalism (Swislocki 2009: 2). Coe has argued (1994: 2) that world studies concerning cooking have contributed to our knowledge of ingredients and their use. The means to process them is also a central issue, and without doubt culinary technology plays a central role. In this chapter, I address Yucatecan Maya culinary technology through a diachronic, multidisciplinary approach that reviews archaeological, historic, and ethno-archaeological data. I emphasize two forms of culinary technology in domestic contexts—grinding and cooking—by analyzing two interrelated dimensions: First, by looking at the different forms of grinding stones (*metates* and *molcajetes*) and cooking tools (pit ovens, hearths) as the result of a wide variety of culinary processes, each with different contextual meanings; and second, by examining the symbolic component of culinary technology. Most of the information I discuss here is the product of ethno-archeological research undertaken during 2012 and 2013 in the village of San Antonio Sihó, Yucatán.

Yucatán, located in the east of the Gulf of Mexico, is far from being homogeneous. Today, *Lo Yucateco* encompasses a mixture of Maya and Spanish populations although its population has been augmented over time with the arrival of groups from around the world, principally from other Mexican regions, but also from Africa, the Middle East (mainly Lebanon), and China and Korea (Ayora-Diaz and Vargas-Cetina 2010). As a result, cooking practices reveal a diversity of recipes, flavors, and meanings that vary according to where they are produced, in regional subdivisions, on the coasts, or in inland cities and small towns.

In Yucatán, speakers of the Maya language inhabit (although not exclusively) rural communities, and even when it is relatively difficult to define what Maya is (which is not the purpose of this chapter), it is possible to argue that language is one of the cultural traits that have remained, with varying degrees of change, from pre-Columbian times. Mayan food has been a topic of interest since the very beginning of the Spanish colonization. The *Relaciones Historico Geográficas de la Gobernación de Yucatán* (Garza et al. 2008) includes descriptions of the first *encomenderos* and reveals the wide variety of available plants and animals that were consumed by the natives, providing details about specific meals and beverages. *Relación de las Cosas de Yucatán*, written by Bishop Diego de Landa in the sixteenth century, includes exhaustive descriptions of ingredients and forms of cooking. Other historical sources allow us to identify some kitchen implements, gender roles, and the consumption of specific meals during specific occasions (Farris 2012; Landa 1986; Roys 1972). Dictionaries can be revealing too: for example, *El Diccionario Maya Cordemex* (Barrera Vásquez et al. 1995 [1980]) includes invaluable and detailed information about ingredients, their combinations, their preparation, and cooking.

During the twentieth and twenty-first centuries, cultural anthropologists and ethno-archaeologists have paid attention to the use of kitchen space and furniture, cooking technology, and foodstuff for both sacred and profane occasions (Redfield 1977 [1946]; Trujillo 1977). Contemporary information about pre-Columbian Maya food traditions comes from different sources: excavations and archaeological analyses of houses and artifacts (Beaudry-Corbett, Simmons and Tucker 2002; Calvin 2002; Cobos et al. 2002; Fernández-Souza 2010; Götz 2005; Toscano Hernández et al. 2011); zoo-archaeological studies have analyzed the menu, preparation, and consumption of animals (Götz 2010, 2011); paleo-ethnobotanical analyses use pollen, starch, phytolites, carbonized seeds, and rinds to reconstruct a vegetal diet (Matos Llanes

2014; Lentz 1999); chemical techniques such as High-Performance Liquid Chromatography (HPLC) with Thermospray MS, Atmospheric Pressure Chemical Ionization and Gas Chromatography–Mass Spectrometry (Henderson et al. 2007; Hurst 2006) are employed to identify chemical signatures, such as the cacao theobromine; pre-Columbian paintings, carvings, and texts depict pottery on walls and other media (McNeil 2006; García Barrios and Carrasco 2008; Kettunen and Helmke 2010; Stuart 2006; Taube 1989); and bio-archaeological studies, including paleopathology, conduct bone chemistry and other analyses (Tiesler and Cucina 2010; White 1999; White et al. 2001).

Over the course of three millennia, ancient Mesoamerican societies and cultures have experienced numerous transformations but authors such as López Austin (2001) stress the existence of a number of practices that constitute what he names *el núcleo duro,* a cultural hard core, that still can be found in many Mexican regions. Corn-based subsistence and ancestral food technology are two examples of these practices. As I discuss below, centuries-old ways of grinding and cooking are still alive thanks to a very old, rich and delicious tradition.

Maya households and kitchen spaces: Archaeological and ethno-archaeological data

Archaeological research conducted in the Mayan Area shows that pre-Columbian house structures were distributed around an open area or *patio;* many of the daily activities were performed outdoors, and there is evidence of grinding both inside kitchen structures and outside on the *patio*. For example, at Sihó, Yucatán, a Maya site in which the main occupation dates to the Classic Period (600–900/100 AD), grinding stones are distributed in *patios* in front of or at the sides of buildings (Cobos et al. 2002; Fernández Souza 2010; Fernández Souza, Toscano Hernández, and Zimmermann 2014). Spot-test chemical analyses practiced in Group 5D72, one of the housing compounds at this site (Matos 2014), showed elevated concentrations of phosphates, carbohydrates, and proteins in the *patio* area, where grinding stones also were found, suggesting food preparation and/or consumption. Outdoor grinding stones have been found close to both palace-type structures and small houses, suggesting that grinding maize outdoors was a common practice, regardless of a household's socioeconomic status (Fernández Souza 2010; Götz 2005; Matos 2014). In Kabah, Yucatán, another Maya Classic Period site, Toscano et al. (2011) reported a large number of grinding stones located in a low terrace situated next to a Maya palace group that connected through stairways. Chemical spot-tests suggest the existence of outdoor fire pits, indicating that people were both grinding and cooking in this area; additionally, two rooms located on the terrace are thought to be storage spaces (Fernández Souza, Toscano Hernández, and Zimmermann 2014; Toscano Hernández et al. 2011). Undoubtedly one of the richest archaeological contexts was located in Joya de Cerén, El Salvador. This site was covered by volcanic ash and thus amazingly well preserved. Excavations at Joya de Cerén have shown probable kitchens, such as Structure 11 in *Solar* 1, where archaeologists found a three-stone hearth (Beaudry-Corbett, Simmons and Tucker 2002), and Structure 16, considered to be the kitchen at *Solar* 3, where charcoal and a large river stone also suggested a hearth (Calvin 2002).

During the first half of twentieth century, Redfield (1977 [1946]: 15) stated that the basic implements within the Maya house were distributed around the hearth, which comprised three ordinary stones; next to it, there was a low round wooden table on which to prepare corn *tortillas* and, as part of the kitchen utensils, a metal mill and a grinding stone were placed on a bench to process fine ingredients. Similarly, Trujillo (1977: 142), in her description of the Maya houses of the Yucatecan henequen hacienda plantation Kankirixche, includes the wooden bench for the grinding stone as part of kitchen furniture.

In contemporary Yucatecan rural communities, food preparation may be done both inside and outside, and a family may have a kitchen—a structure separated from the house—with one hearth, plus one additional hearth or two in the *patio*. At San Antonio Sihó, Yucatán, a community located in the vicinity of the archaeological site of Sihó, it is common for families to do some food preparation in the kitchen, but other parts of the process may be done outside, such as cleaning the maze, washing dishes, or plucking edible farmyard birds. As noted above, many kitchens are semi-open constructions with a three-stone hearth and/or a cement hearth (and sometimes a modern gas stove too), with low tables and benches to make *tortillas*, and taller plastic or wooden tables and chairs to eat and prepare foodstuffs. Kitchen implements are stored on shelves or in bags suspended from the roof. Some kitchens are completely open on one of their sides, so that the cooking and eating space is a continuum between the house and *patio*. Grinding is usually done in the kitchen area, but, as discussed above, it can be done either inside or outside, in the *patio* or next to the kitchen, protected by an outward straw roof. Since trees are ubiquitous in Maya *patios*, their shade and coolness may entice the family to prepare dishes and eat outdoors, especially during the hottest days of the Yucatecan spring and summer. Cooking may be done in the three-stone or cement hearths; in addition, the house may include a *pib* or pit oven, a hole that is filled with red-hot stones to roast and bake special dishes. These ovens are dug in the *patio*, in elevated areas usually immune to floods. Homemakers commonly raise domestic animals (mainly hens and turkeys) and grow edible garden plants, such as chilis, tomatoes, herbs, Seville oranges, and other fruits (mamey, pitahaya, *ciricote*, avocado, or *caimito* or star apple). This practice has been reported since early colonial times (Garza et al. 2008; Landa 1986) and continues to make many rural Maya houses beautiful.

The form and function of grinding stones

The grinding process is very important for Yucatecan Maya cooking; grinding stones were, and still are, useful to mill foodstuffs, but they have also been employed to grind ceramic materials and minerals for pigments (Götz 2005: 72). There are examples of very early grinding stones in pre-Columbian Mesoamerica; for example, Acosta Ochoa et al. (2013: 539) report starch grains of *Zea* sp. (*teosinte*, an ancestor of maize) on a grinding stone found in the cave of Santa Marta, Chiapas, México, dating back to 9800 BP. Götz (2005) points out that the morphology of a grinding or milling stone (named *metate* or *ka'* following Nauatl or Maya words, respectively) is the result of a variety of factors such as chronology, a user's socio-economic status, specific functions, and specific group traditions. Based on his analyses of *metates* from Mayan archaeological sites like Dzibilchaltun, Komchen, Misnay, and Kaua, Götz concludes that important variations result from chronological differences: for example, pre-Classic *metates* are generally bigger and more irregular than those from the Classic period. Yucatecan pre-Classic and Classic *metates* used to be legless rectangular stones deposited directly on the ground or over smaller supporting stones although, at Joya Cerén, archaeologists found a *metate* supported by a wooden holder. Stuart (2014) has proposed an alternative reading for the glyphic name of a site that was previously known as *chi-witz*; he suggests that the name was actually *chi-ka* or *chi-cha*, meaning "maguey milling" or "the place of maguey milling." According to Stuart, the glyph shows a legless *metate* with a stone support. Over time, Yucatecan *metates* became three-legged, just as they are today.

The importance of the *metate* (*ka'*, in Yucatec Maya [Barrera Vázquez et al. 1995: 277]) for daily kitchen activities may be recognized both in archaeological contexts and historical sources. In Kabah's kitchen context, next to the palace complex, Toscano et al. (2011) reported around thirty legless grinding stones; three domestic structures, located at the center of the aforementioned Sihó site, had between five and eight *metates* each (Cobos et al. 2002; Fernández Souza 2010; Matos 2014). Götz (2005: 93) found something similar in Dzibilchaltun, where he reports one to eight grinding stones in each of the households' multiple platforms.

In the sixteenth century Friar Diego de Landa (1986: 43) described the preparation of corn meals and beverages: "Maize is the main nourishment, from which they prepare diverse delicacies and beverages [. . .] and the

Indians soak corn in water and lime the night before, and in the morning it is soft and half-cooked [. . .]; they mill it on stones and give the half-milled [maize] to the workers, walkers and sailors . . .'[1] He also describes a frothy beverage prepared with toasted and ground corn and cacao (Landa 1986: 43). Other colonial documents such as the *Relación de Sotuta y Tibolom* and the *Relación de Tabi and Chunhuhub* (Garza et al. 2008: 148, 165) mention that the natives consumed a beverage similar to *poleadas* (also mentioned by Landa 1986: 43), a Spanish word translated in the Motul Dictionary as "*sa'*: *atol, que son gachas, puches de masa de maiz*" [*sa': atol,* which is a porridge made of maize dough] (Barrera Vázquez et al. 1995: 707). The *Relación de Hocabá* reports the belief that the natives were healthier in former times because they used to drink a "wine" made out of water, honey, ground maize, and roots (Garza et al. 2008: 134).

During the first half of twentieth century Redfield (1977 [1946]: 15) observed the coexistence of *metates* and metallic mills in the same kitchen. He proposed that metallic mills were in use "even in the most remote hamlets, but most families use the *metate* to prepare fine dishes." This coexistence is found today in many rural Yucatecan houses; some families own a three legged *ka'*, a metal mill, *and* an electric blender. In villages there are frequently one or more commercial *molino y tortillería,* establishments that mill corn and sell machine-manufactured maize *tortillas.* At Sihó, women can choose to either bring their *nixtamal* (corn soaked in water and lime) to the mill and prepare hand-made *tortillas* at home, or buy them at the *tortillería*. However, *metates* are no longer used to grind corn: today, their use is restricted to grinding *achiote* (*bixa orellana*) and other condiments. Some people have told me that *frijol colado* (strained black beans) was more delicious when it was prepared over a *metate*. One of the ladies at Sihó explained that beans tasted better when they were ground in the stone *ka',* but more was wasted than when using the blender; the blender allows her to make the most of the food. This feeling of nostalgia can be noted around a variety of cooking processes; for Yucatecan people (regardless of city or small village provenience), hand-made *tortillas* will always be more delicious than the machine-made variety. Nevertheless, for special occasions, such as the Day of the Dead on October 31 and November 1, traditional *mucbil pollos* recipes (large *tamales* baked in *pibs*, ground ovens) require *achiote*, which is still ground in three-legged *metates*.

A number of Maya words related to grinding or milling underline the importance and variety of this action. For example, the *Diccionario Maya* (Barrera Vázquez et al. 1995) includes entries such as *nil* ("*moler bien*" or "milling well"); *tikin huch'* ("*moler en seco*" or "dry milling"), and *tan chukwa'* ("*chocolate molido con masa y especias*," chocolate ground with corn dough and spices), to mention just a few (Barrera Vázquez et al. 1995: 571, 793, 772).

In addition to the *ka'* there is another presumably ancestral instrument that is quite ubiquitous in Maya Yucatecan kitchens—the mortar or *molcajete*. Some Maya words for mortars are *ch'en tun* (which means "stone mortar," Barrera Vázquez et al. 1995, 133) and *likil mux*. With their *k'utub* (the mortar's hand), mortars are useful to *k'ut* (to mash chili, mustard, or parsley with a little water or juice, Barrera Vázquez et al. 1995: 424), quite an important procedure for the preparation of chili sauces.

The *Recetario Maya del Estado de Yucatán* (Maldonado Castro 2000) is a cookbook that records a number of southern Yucatán dishes that clearly show the central role of grinding. For example, *óonsikil chaay* ingredients are *chaya* leaves, toasted pumpkin seeds, chili, *achiote*, tomatoes, corn dough, local plums, salt, pepper, and chives. For this dish, pumpkin seeds must be finely milled along with the chili. To prepare *tóoncha' ch'oom xpaapa' ts'uul* (boiled egg-like tacos covered with a pumpkin seed paste and tomato sauce), the cookbook indicates that it is necessary to toast, grind, and strain the seeds to obtain a very fine texture, after which roasted tomato is mixed in with the chili (Maldonado Castro 2000: 49, 51). At Sihó, Doña Rosa, one of our subjects, prepares *óon sikil bu'ul* with care and patience. For this dish—made of a mixture of toasted pumpkin seeds, *achiote*, chili, and maize dough which is mixed into a boiling black bean broth—she grinds nixtamalized corn (at the village's *molino y tortillería* or in her own electric mill) and *achiote* in her *metate*, pumpkin seeds in her iron mill, and chili in a plastic mortar. She also has to strain the mixture of ingredients several times before she pours it into the boiling pot of beans. Chocolate is prepared at home for special occasions, particularly for the Day of the Dead. Like many other Sihó women, Doña Rosa grinds toasted cacao seeds in an iron mill. She grinds the seeds as many times as necessary to obtain a fine and delicate texture to prepare chocolate bars that, whipped with hot water, honey, or sugar, result in a delicious and frothy traditional beverage.

Cooking in diversity: The *pib* and the *k'oben*

The three-stone hearth or *k'oben* is used for everyday cooking. It is defined as "fire stones over which they put the pot" and "the three stones forming a hearth"; similarly, *k'obenil k'ak'* means "the three stones of the Maya hearth, which hold the pot and the griddle (*comal*)" (Barrera Vázquez et al. 1995: 406). As noted above, the Classic Maya site Joya de Cerén presents two features that have been suggested to be *k'obenob*. Redfield (1977 [1946]) wrote that cooking in a pot and roasting over the grill were culinary art's main methods among the Maya, adding the *pib*. Nevertheless, there were, and are, several other ways of cooking. Examining dictionary entries (Barrera Vázquez et al.

1995), one finds the meaning for cooking in the embers (*pem chuk*), to cook without water or broth (*nakmal*), to cook in a pot (*tubchak*), to roast and toast (*op'*), to fry (*tsah*), and to fry without grease (*tikin sah*). All of these procedures can be performed over a three-stone hearth. In addition, the subterranean *pib* oven provides a style of cooking that consists of inserting red-hot stones (thermoliths) into the casserole containing the stew. This latter procedure is known in Yucatecan Maya as *toccel*. A very similar one is known to have existed in Mesoamerica: Clark, Pye and Gosser (2007) mention the presence of thermoliths during the early Formative Period (1900–1000 BC) in the Mazatan region of Chiapas, Mexico, suggesting it is a continuation of culinary practices originating in the distant past.

In contemporary rural Yucatán, in a similar vein to grinding processes, modern and traditional cooking technologies coexist without conflict. At Sihó, Doña Lucy and her mother, Doña Imelda, frequently cook together, although they live in separate houses. There is a gas stove at Doña Imelda's house, but as gas is quite expensive they prefer to use the *k'oben*. They also own a cement hearth and an iron *barbecue*. From an archaeological standpoint, this is a fascinating scenario, because we find at least five millennia of cooking technologies deployed at the same place at the same time. At Don Geydi's home, a broken cement hearth was repaired simply by adding two stones; as a result, this appliance is now half a hearth and half a *k'oben*. On the opposite side of the kitchen, Don Geydi's wife cooks in another *k'oben* by transforming a fan rack into a grill.

More than one cooking technique may be used for the same recipe. For example, to prepare the aforesaid *tóoncha' ch'oom xpaapa' ts'uul*, one must toast pumpkin seeds, boil eggs and *chaya* leaves, and roast tomatoes (Maldonado Castro 2000: 51). Maize beverages and meals are also cooked in a wide variety of ways: *tamales* may be steam-cooked; *tortillas* (*waaj*) are made on the *comal* (Mayan *xamach* or griddle); thick *tortillas* (*pem chuk*) are cooked over embers; *mucbilpollos* are baked in the pit oven; *panuchos* and *salbutes* are fried *tortillas* with and without beans, respectively, and topped with poultry, onion, and tomato slices. It is important to stress that as a consequence of this wide range of cooking techniques, a number of core ingredients can be used in different ways to spawn a rich culinary panorama.

In the land of gods: Ritual and gender practices regarding/pertaining to kitchen technology

Maldonado Castro (2000: 24) classifies contemporary Maya food into three groups:

1. everyday plain food, mostly cooked with seasonally available ingredients;
2. festive food, prepared for special occasions like birthdays, weddings, baptisms, and others;
3. ceremonial food destined to be offered and consumed during rituals such as the *cha chaac* (a ceremony performed to pray for rain) or the *wajil kool*, a ritual offering to the lords of the *milpa* to thank them for the harvest.

As suggested by Fernández-Souza, Novelo Pérez and Cu Pérez (2014), everyday food in western Yucatecan villages includes many non-local ingredients, such as rice, wheat flour, bread and pasta, sugar, oranges, limes, and onions, some of which were introduced shortly after the Spanish conquest. Other meals and beverages have gained popularity during the last century, like the cola sodas that are now ubiquitous in every house and at every meal. On the other hand, for ceremonial food, especially that which is offered during forest and *milpa* rituals, local ingredients like maize, *achiote,* and cacao strongly dominate the culinary panorama. There is evidence underscoring the importance of food in pre-Columbian Maya ceremonies. For example, Nájera Coronado (2012: 137) refers to page 25a of the Dresden Codex, where it is found that, during the calendar *wayeb* ceremonies, cacao was offered to god *K'awiil,* since this beverage was considered to be his nourishment. On the other hand, for the ancient Maya, death constituted a long and grievous path for the deceased, and as a consequence, the relatives placed corn in his or her mouth and pottery plates and vases containing meals and beverages were deposited in or on the graves (Eberl 2000; Tiesler and Cucina 2010).

For colonial Maya society, Farris (2012: 419) suggests that the role of food was reasonably important during religious festivities, especially those dedicated to patron saints. Every celebration, including funerals, weddings, house constructions and so on, involved a communal banquet. According to Farris (2012: 419), feasts gave common people the opportunity to taste food that was not normally available to them, such as pork, beef, plenty of lard, expensive condiments and alcoholic beverages (Farris 2012: 419, 420).

Metates and the *k'oben* still find ritual uses. During pre-Columbian times grinding stones and or their hands were deposited, presumably as offerings, at the corners and centers of both religious and non religious structures. Such is the case of El Templo de las Grandes Mesas, at Chichén Itzá and Structure 8 at Isla Cerritos (Castillo Borges 1998; Fernández Souza 2008). Nowadays, at Sihó, it is still common to bury a *metate* hand of the *ka'* at the center or the corners of new houses, or in newly constructed rooms. People say that this is a *k'ex*, an exchange with the soul of the house to protect family members, especially children. On the other hand, the three-stone *k'oben* is also the center of a gendered ceremony: when a baby boy is born, his umbilical cord is taken into the forest, so that he is not frightened about hunting or planting in later life. If a baby girl is born, the umbilical cord is deposited in the ashes of the hearth, because her parents and society expect her to become a housekeeper who stays home and takes good care of her family.

Concluding remarks

Many things are rapidly changing in the contemporary Maya lands. As mentioned by Maldonado Castro (2000: 28), canned products, industrial food, imported condiments, and colorants are modifying the culinary panorama of Yucatán. The feminine figure has been central to rural cuisine, but in modern communities young women are not as interested as their female forbears in learning how to cook (Maldonado Castro 2000: 36). Electrical appliances are replacing old implements, and tools, such as clay griddles (*comal* or *xamach*) have been long substituted by their metal equivalents. However, many traditional features firmly coexist with modern technology. Most families at Sihó preserve the *k'oben* largely due to the high cost of gas for cooking. The *pib* or pit oven plays a fundamental role for the preparation meals and other festivities for the Day of the Dead, and is still accorded high symbolic importance in rural and urban Yucatán; however, in the city, traditional *mucbilpollos* are baked in conventional ovens or taken to bakeries. Without pretending that Maya culinary technology must be kept alive as a museum relic, or that women should spend hours grinding corn in a *metate,* I consider it important to discuss and consider traditional ingredients and culinary practices. On the one hand, most of the aforementioned dishes are highly nutritious and, being local, they may be more sustainable than other contemporary practices. On the other hand, and as noted, food plays a role in reinforcing a sense of community, sociability, and identity ties. Some aspects of Mayan food and cuisine technology have survived for at least 3,000 years; by studying and exploring their history, development, and transformation, we can contribute to our knowledge of cultural diversity, change, and continuity. Finally, in a world where each day it is harder to feed the ever-growing human population, subsistence strategies and technologies from the past may present us with useful discussion topics to think about for our future.

Note

1 Unless noted, all translations are the author's.

References

Acosta Ochoa, G., Pérez Martínez, P., and Rivera González, I. I. (2013), "Metodología para el estudio del procesamiento de plantas en sociedades cazadoras-recolectoras: un estudio de caso." *Boletim do Museu Paraense Emílio Goeldi Ciências Humanas* 8(3): 535–330.

Ayora-Diaz, S. I. (2012), *Foodscapes, Foodfields and Identities in Yucatán*. Amsterdam: CEDLA / New York: Berghahn.

Ayora-Diaz, S. I., and Vargas-Cetina, G. (2010), "Introducción. Antropología y la Imaginación de lo Yucateco." Pp. 11–32 in S. I. Ayora-Diaz and G. Vargas-Cetina (eds), *Representaciones Culturales: Imágenes e Imaginación de lo Yucateco*. Mérida: Universidad Autónoma de Yucatán, Mérida.

Barrera Vásquez, A., Bastarrachea Manzano, J. R., Brito Sansores, W., Vermont Salas, R., Dzul Góngora, D., and Dzul Poot, D. (eds) (1995 [1980]), *Diccionario maya cordemex: maya–español, español–maya*. Mérida: Ediciones Cordemex.

Beaudry-Corbett, M., Simmons, S. E., and Tucker, D. (2002), "Ancient Home and Garden: The View from Household 1 at Cerén." Pp. 45–57 in Sheets, P. (ed.), *Before the Volcano Erupted*. Austin: University of Texas Press.

Calvin, I. (2002), "Structure 16: The Kitchen of Household 3." Pp. 45–57 in Sheets, P. (ed.), *Before the Volcano Erupted*. Austin: University of Texas Press.

Castillo Borges, V. R. (1998), "Liberación y restauración de la Estructura 2D7 o Templo de las Grandes Mesas de Chichén Itzá." B.A. thesis. Mérida: Universidad Autónoma de Yucatán.

Clark, J., Pye, M. E., and Gosser, D. (2007), "Thermolithics and Corn Dependency in Mesoamerica." Pp. 23–61 in L. Lowe and M. Pye (eds), *Papers in Honor or Gareth W. Lowe*. Provo, Utah: New World Archaeology Foundation, Brigham Young University.

Cobos, R., Fernández Souza L., Tiesler, V., Zabala, P., Inurreta, A., Peniche, N., Vázquez, M. L., and Pozuelo, D. (2002), "El Surgimiento de la Civilización en el Occidente de Yucatán: los Orígenes de la Complejidad Social en Sihó." Fieldwork report 2001 presented at the Consejo de Arqueología del INAH. Mérida.

Coe, S. (1994), *America's First Cuisines*. Austin: University of Texas Press.

Eberl, M. (2000), "La muerte y las concepciones del alma." Pp. 311–318 in N. Grube (ed.), *Los mayas: Una civilización milenaria*. Cologne: Könemann.

Farris, N. (2012 [1984]) *La sociedad Maya bajo el dominio colonial*. Mexico City: CONACULTA.

Fernández-Souza, L. (2008), "Los dioses que nunca se fueron: Ceremonias domésticas en el Norte de la Península de Yucatán." Pp. 1029–1040 in J. P. Laporte, B. Arroyo, and H. Mejía (eds), *XXI Simposio de Investigaciones Arqueológicas de Guatemala 2007*. Guatemala: Museo Nacional de Arqueología y Etnología.

Fernández-Souza, L. (2010), "Grupos domésticos y espacios habitacionales en las Tierras Bajas Mayas durante el Período Clásico." Doctoral Thesis, Hamburg University, available online at http://ediss.sub.unihamburg.de/volltexte/2010/4512/pdf/Dissertation.pdf.

Fernández-Souza, L., Novelo Pérez, M. J., and Cu Pérez, A. (2014), "La cocina rural contemporánea: aproximación etnoarqueológica a la gastronomía de tres comunidades del occidente de Yucatán." Pp.73–90 in *Los Investigadores de la Cultura Maya 22, Vol. 2*. Campeche: Universidad Autónoma de Campeche.

Fernández Souza, L., Toscano, L., and Zimmermann, M. (2014), "De maíz y de cacao: aproximación a la cocina de las élites mayas en tiempos prehispánicos." Pp. 107–130 in S. I. Ayora-Diaz and G. Vargas-Cetina (eds), *Estética y Poder en la Ciencia y la Tecnología. Acercamientos Multidisciplinarios*. Mérida: Universidad Autónoma de Yucatán.

García Barrios, A. and Carrasco, R. (2008), "Una aproximación a los estilos pictóricos de la Pirámide de las Pinturas en la Acrópolis Chiik Nahb de Calakmul." Pp. 848–867 in J. P. Laporte, B. Arroyo, and H. Mejía (eds), *XX Simposio de Investigaciones Arqueológicas de Guatemala, 2007*. Guatemala: Museo Nacional de Arqueología y Etnología.

Garza, M., Izquierdo, A. L., León, M. del C. and Figueroa, T. (2008), *Relaciones Histórico-geográficas de la Gobernación de Yucatán*. Mexico City: Universidad Nacional Autónoma de México.

Götz, C. (2005), "Una tipología de los metates prehispánicos de Yucatán." *Ketzalcalli* 2: 70–99.

Götz, C. (2010), "Una mirada zooarqueológica a los modos alimenticios de los mayas de las tierras bajas del Norte." Pp. 89–109 in H. Hernández Álvarez, H. and C. Pool Cab (eds), *Identidades y cultura material en la región maya*. Mérida: Universidad Autónoma de Yucatán.

Götz, C. (2011), "Diferencias socioeconómicas en el uso de animales vertebrados en las tierras bajas mayas del norte." Pp. 45–65 in R. Cobos, and L. Fernández-Souza (eds), *Vida cotidiana de los antiguos mayas del norte de la Península de Yucatán*. Mérida: Universidad Autónoma de Yucatán.

Henderson, J. S., Joyce, R. A., Hall, G. R., Hurst, W. J. and McGovern, P. E. (2007), "Chemical and Archaeological Evidence for the Earliest Cacao Beverages." *PNAS* 48(27):18937–18940.

Hurst, J. (2006), "The Determination of Cacao in Samples of Archaeological Interest." Pp. 105–113 in C. L. McNeil (ed.), *Chocolate in Mesoamerica*. Gainesville: University Press of Florida.

Juárez López, J. L. (2008), *Nacionalismo Culinario. La Cocina Mexicana en el Siglo XX*. Mexico City: CONACULTA.

Kettunen, H. and Helmke, C. (2010), *Introducción a los Jeroglíficos Mayas*. Available online at http://www.mesoweb.com/es/recursos/intro/JM2010.pdf

Landa, D. de (1986), *Relación de las Cosas de Yucatán*. Mérida: Consejo Editorial de Yucatán A.C.

Lentz, D. L. (1999), "Plant Resources of the Ancient Maya. The Paleoethnobotanical Evidence." Pp. 3–18 in C. White (ed.) *Reconstructing the Ancient Maya Diet*. Salt Lake City: The University of Utah Press.

López Austin, A. (2001), "El núcleo duro, la cosmovisión y la tradición mesoamericana." Pp. 47–65 in J. Broda and F. Báez-Jorge (eds), *Cosmovisión, ritual e identidad de los pueblos indígenas de México*. Mexico City: CONACULTA-FCE.

Maldonado Castro, R. (ed.) (2000), *Recetario Maya del Estado de Yucatán*. Mexico City: CONACULTA, Culturas Populares, Instituto de Cultura de Yucatán.

Matos Llanes, C. M. (2014), "Alimentación vegetal y áreas de actividad en la unidad habitacional 5D72 de Sihó Yucatán. Etnoarqueología, análisis químico de suelos y paleoetnobotánica como herramientas de aproximación." B.A. thesis. Mérida: Universidad Autónoma de Yucatán.

Nájera Coronado, M. I. (2012), "El Mono y el Cacao: la búsqueda de un mito a través del Grupo de la Serie Inicial de Chichén Itzá." *Estudios de Cultura Maya* XXXIX: 136–167.

Redfield, R. (1977 [1946]), "Los Mayas actuales de la Península Yucatanense." Pp. 7–30 in L. Hoyos Villanueva, R. Ruz Menéndez, R. Irigoyen Rossado, and H. Lara y Lara (eds), *Enciclopedia Yucatanente. Vol. VI*. Mexico City: Gobierno del Estado de Yucatán.

Roys, R. (1972), *The Indian Background of Ancient Yucatán*. Norman: University of Oklahoma Press.

Scholliers, P. and Claflin, K. W. (2012), "Introduction: Surveying Global Food Historiography." Pp. 1–8 in K. W. Claflin and P. Scholliers (eds), *Writing Food History: A Global Perspective*. London and New York: Berg.

Swislocki, M. (2009), *Culinary Nostalgia. Regional Food Culture and the Urban Experience in Shanghai*. Stanford: Stanford University Press.

Stuart, D. (2006), "The Language of Chocolate: References to Cacao on Classic Maya Drinking Vessels." Pp. 184–201 in C. L. McNeil (ed.), *Chocolate in Mesoamerica*. Gainesville: University Press of Florida.

Stuart, D. (2014), "A Possible Sign for Metate." In *Maya Decipherment*. https://decipherment.wordpress.com/2014/02/04/a-possible-sign-for-metate/

Taube, K. (1989), "The Corn *Tamale* in Classic Maya Diet, Epigraphy and Art." *American Antiquity* 54 (1):31–51.

Tiesler, V. and Cucina, A. (2010) "Descubriendo los secretos de los huesos." Pp. 49–60 in Fernández Souza, L. (ed.), *En los antiguos reinos del jaguar*. Mérida: SEP, Gobierno del Estado de Yucatán.

Toscano Hernández, L., Jiménez, R., Novelo, G., Ortegón, D., Duarte, A., Castro, K., Marengo, N., García, O., and Cruz, J. (2011), Proyecto Investigación y Restauración Arquitectónica en Kabah, Yucatán. Fieldwork Report 2010. Document presented at the Archivo Técnico del INAH, Mexico City.

Trujillo, N. (1977), "El Maya de las Haciendas Henequeneras." Pp. 133–171 in L. Hoyos Villanueva, R. Ruz Menéndez, R. Irigoyen Rossado, and H. Lara y Lara (eds), *Enciclopedia Yucatanente Tomo VI*. Mexico City: Gobierno del Estado de Yucatán.

White, C. D. (1999), "Introduction." Pp. ix–xxvii in White, C. D. (ed.), *Reconstructing Ancient Maya Diet*. Salt Lake City: University of Utah Press.

White, C. D., Pendergast, D. M., Longstaffe, F. J., and Law, K. R. (2001), "Social Complexity and Food Systems at Altun Ha, Belize: The Isotopic Evidence." *Latin American Antiquity* 12(4):371–393.

2

Technology and culinary affectivity among the Ch'orti' Maya of Eastern Guatemala

Julián López García, Universidad Nacional de Educación a Distancia, Madrid; Lorenzo Mariano Juárez, Universidad de Extremadura

Introduction: "Appropriate" technology as an expression of neocolonialism

When we began our field work among the Ch'orti' Maya of Eastern Guatemala during the second half of the 1980s, approaches to so-called "appropriate technology" for change in indigenous communities in that location were in vogue. They were rooted in an idea, inherited from 1970s' developmentalism, which justified attempts to modify or eliminate supposedly inappropriate (or inadequate) autochthonous technologies. Various governmental and nongovernmental programs were working toward these aims, and the local radio station, Radio Ch'orti', featured a program called "Appropriate Technology," in which local expert technicians promoted changes in agricultural techniques and procedures.

At that time, the emphasis on change to support the adoption of "appropriate technology" was focused primarily on matters relating to agricultural production. Very often, technicians singled out the ineffective use of "slash and burn" agriculture in the cornfields, the inefficiency of seed-selection processes, and the counterproductive effects of crop and land rotation. Accordingly, they recommended the introduction of new varieties of high-yield corn, of a system of terraces and natural barriers (used to grow *gandul*, *izote*, pitahaya and pineapple that were promoted to prevent soil washing); and new and varied crops that were supposedly better suited to the land, more profitable, and with a higher nutritional value. Advocates of these new technologies were found celebrating what they saw as a major victory: the successful introduction of sorghum, which began to colonize cornfields and displace corn during the second harvest. Rural "promoters" (the name given to the last links in this chain of change) were overjoyed by their triumph, as it had been hard work to overcome the indigenous populations' resistance to changing their traditional agricultural system. That system was based on the preeminence of corn, whose value to the Ch'orti' culture—and for the Maya people in general—was much greater than mere nutrition. These promoters praised the efficiency of sorghum, its nutritional values, its easy adaptation to the region's soil, and high crop yield. However, they also suggested that natives' resistance had been defeated and that sorghum had been allowed to share space in the cornfield not for nutritional and financial reasons, but for something rather symbolic and ideological: instead of calling the new product "sorghum," the natives christened it *maicillo*.[1] The perceived victory was anything but, however. In essence, yet another trick had been played on the indigenous populations to steer change. The suggested ideological connection was clear: by using the diminutive *maicillo*, promoters suggested that the new grain was "small corn"—different in maturation (it took three months longer to mature), different in color (darker than the usual yellow or white), and with a "slightly" different flavor, but still basically corn.

Maicillo became part of the region and, together with corn, constituted the raw material for making *tortillas*. This did not mean that it was accepted, though, or accorded the same value. As Guadalupe, one of our subjects, recognized at the beginning of the 1990s, *maicillo* had some characteristics that were obviously negative compared to corn. For example, she said, it has an unpleasant smell ("*maicillo tortillas*, yuk! They reek like a pig sty") and goes hard rapidly and easily ("even though you make a *maicillo tortilla* and pull it off the *comal*, it hardens right up in your hands as if it were three days old"). Despite celebratory accounts from the promoters and satisfaction about modernity's triumph in the shape of grain diversification, in fact, the accounts from our subjects told a very different story: that sorghum, as much as it may be *maicillo*, could in no way compete with yellow or white corn. The feeling that *maicillo* is an imposed grain is very present in the memory of rural people, and very often in their testimonies they allude to a "ruling," a type of order issued by those in power that has been locally accepted. Thus the new grain was not adopted out of pleasure or conviction, but rather as the result of a mandate and, as we know, there is nothing more disagreeable to culinary enjoyment than the imposition of change.

We began by pointing out the paucity of dialogue between the apostles of "appropriate" technology and those who follow their traditions in order to argue that an adoption of culinary technologies based on a lack of sincere dialogue is doomed to fail, even when it competes with what modernizers see as technological simplicity and primitivism. With this premise, and cognizant of the many who have set out to "modernize" the countryside, in this chapter we focus on the debate over the reach and limitations of technological change inside the kitchens of the Ch'orti'.

Culinary technology for corn *tortillas*: The *metate* and the *comal*

For the Ch'orti' people, as for the rest of the Maya, the corn *tortilla* constitutes the foundation of nutritional and symbolic subsistence. The making of *tortillas* has oriented many aspects of Ch'orti' life, from understanding the underpinnings of the marriage contract to gender distinctions; and it goes further to establish the roots of ethnic identity (Mariano Juárez 2013) or the logic of the inhabited topography.

The *tortilla* epitomizes the essence of real food. In Ch'orti' language, no words exist for "meal" or "to eat"; normally these are both referred to as *ni pa'* ("my *tortillas*"). *Tortillas* are the food with greatest emotional involvement since they both nourish the body and sustain the soul. They are the only food that satisfies hunger (Mariano Juárez 2013) but, moreover, transmits the idea of humanity: "real men eat *tortillas*." Nevertheless, as we have pointed out in other publications (López García 2000: 366–368), not all *tortillas* are created equal. Rather, the spectrum runs from the *tortillas* that are *fiotas* (ugly, imperfect) and completely dissatisfying, to those considered *galanas* (elegant). The former have an unattractive appearance because they are not perfectly rounded, or the grain was not boiled with enough lime, or the dough was inadequately kneaded. Perhaps they are undercooked or burned, or have a strange taste, or even show signs of poor hygiene standards in the women who prepared them, evidenced by the presence of a hair or a piece of straw in the end product. In contrast, a *tortilla galana,* is at once beautiful and good, the outcome of many hours invested in its preparation. The *tortilla galana* highlights the emotional values mobilized during its preparation. Although time could be measured in terms of the hours required for cooking and washing, milling and shaping *tortillas*, there is something more to it, that which in the sociology of emotion is called "emotion work," and that in this case explains the overabundance of diminutives while the *tortillas* are cooked and served. The women display a formal reverence for the cooking process: there are gestural and domestic ritual formalities to be observed during their preparation, and whispers during the work are intended to reprove other women who make fun of *tortillas* they consider *fiotas*. To successfully manufacture a *tortilla* considered *galana,* women need to deploy an arsenal of time and emotions, as well as a technical and emotional mastery in the required procedures to mill the grain, and the movements to shape the *tortilla*. But success also requires the employment of the two basic tools used to cook it: the *metate* (milling stone), and the *comal* (griddle). In a way, the emotional value given to *tortillas* is transferred to these tools, as well as to the palm of the hand that, like any other technical and artistic instrument, produces a beautiful and good *tortilla*. Among the Ch'orti', having a home requires a man who brings the corn home, women with capable hands, and owning a *metate* and a *comal*. In essence, the home constitutes the space in which the most locally important feminine work is done: making corn *tortillas*.

The milling stone and the griddle are used during the process of making *tortillas*. These basic tools are considered important in indigenous kitchens, and their value stems from the fact that their association with the *tortilla* inscribes a unique connotative value. They are objects with a story to tell and onto which important social values are projected; they are tools that turn out to be ideologically different: the *metate* is the most enduring domestic tool, and the *comal* is one of the most ephemeral. In that sense, the *metate* is tied to vital processes, to women's trajectories while in contrast the *comal* is tied to moments in life. The *comal* relates to daily events and to small discontinuities during short periods of time, while the milling stone accompanies transcendental generational developments. One would be tempted to think that it is the radically different raw materials used to make them that facilitate the transmission of different ideas: the fragility of clay in contrast to the durability of stone conveys correlations that refer either to the day-to-day or to an entire lifetime. The meaning is more complex, however. We need to go beyond the durability of the milling stone and the fragility of the clay to examine the attributed meaning of the tools and the biographies constructed around them (López García 2003: 114).

The life of a *comal* partly depends on the *tortillas* cooked on its surface. The presence of *comal tajos* in the home, of broken pieces of *comales*, means that many *tortillas* have been made and many relationships formed; these broken pieces suggest happiness. Broken pieces of *comal* scattered around the house are turned into decorative motifs that embellish the home, indicating a dwelling that is not just happy, but also beautiful. This is an idiosyncratic aesthetic code for decoration in which parts of the building are comprised of the remaining pieces of objects that have been deemed good and desirable, and which have been used and consumed. Thus, on the floor of the house, leaning against the walls, a visitor can always find pieces of *comales*, broken clay pots and crates and, along the exterior and interior walls of the dwelling, egg shells, bird *cacastes* (skeletons), *chumpe* (turkeys) or chickens feathers, armadillo armor, or even pieces of iguana skin; again, things that were part of something good that was consumed. Since Doña Gregoria is already an old woman who only has to make *tortillas* for four people, she has few pieces of *comal* in her house, but any outsider can still recognize pieces belonging to no less than three different *comales*. The first *tajo* belongs to a *comal* that Doña Gregoria bought on St. James' Day "last year" (2001), a small piece she placed in the corner:

> This piece is from a *comalito* [small *comal*] I bought in the Jocotán market on the day of the fair. There were periods in which I did not make *tortillas* in *comales* from Matasano, even though it was out of pleasure that I returned carrying it on my back, because this one is in the same style as those from Pacrén. I did not start using it that day, but rather until *sikín*.[2] Three female millers turned up and we made heaps of *tortillas* . . . it was then when I first used it. Around that time Rosita had stopped making *tortillas*. She was seven months pregnant and it was not good for her to *agarrar calor* [catch heat], let alone to touch the milling stone. This *comalito* broke four months ago . . . last May, and by then I could not make my *tortillas* for the Day of the Cross . . . I first used that other piece over there for the Day of the Cross. That one was brought in from Pacrén. I bought it on April 25 for 25 pesos . . . that *comalito* didn't last long, not even five months, but this *comalito* had to support the boys that planted, of whom we had many this year, and also boys that harvested . . . that [third] small piece over there is from a *comalito* about three years old, almost four. In that *comal* I made *tortillas* for Manuel, who's now deceased . . .

The idea of the warm *comal* is so inviting for the woman that it becomes the metaphor for the ideal home. As they often say in Tunucó, it is best for a woman to be with a man who can keep the palm of her hand warm, and a women's hand is warm when the *comal* is warm; which is to say, the woman will be contented because she is certain she will have enough *tortillas* to eat in the house. We have found that these villagers like bringing home the *comales* in one long procession, in such a way that everyone in the community can see them, and so that it is conveyed that at that specific house the women keep breaking *comales* and making *tortillas*. The fact that women frequently break them suggests that marital relations are, unlike the *comales*, strong.

Generally speaking, most homes have a primary milling stone that is used every day to make dough. This milling stone ideally possesses the right surface coarseness to effectively pulverize the boiled corn grains; usually the family will own another stone mill to grind coffee beans. That second mill could, if thoroughly washed, also be used alongside other mills to grind corn, especially if there is a social event that requires several millers to join forces. Doña Gregoria speaks of the milling stone she uses daily to make dough:

> My mom bought it so that I could start milling with it. I started milling when I was eight [years old]. I milled very little and then, after a few years, I could mill quite a bit. Mothers always first show us how to grind and they grab the stone to refine the dough. My mom used to tell me to watch her so that I could see what it was like to mill and, what it was like to refine the dough. On that stone I also learned how to refine the dough . . . I took it with me when I moved away to Tunucó Arriba with my first partner. There I milled the dough for our wedding invitation and also for the funerals of the children I had with that man. We milled dough [at the funerals] but I just could not, so it was the other visitors who did it. That sickness took hold of me to the point that I couldn't mill any more.[3] I brought [the stone mill] with me when I came back here. I looked for a young lad to bring it because I could not bring it myself; because it's pretty heavy . . . the boy brought it here to this house, which belonged to my mom, who was still here when I came back with it. I'm providing for Don Luis and Agustín with the stone and I also use it to mill the *memela* [*tortilla*] dough for the grandchildren. Cristina, Lupa, María, and now also Rosita[4] all made it through the *totopostes* [eating food made] from that stone . . . It has done its job well. Don't you see how it is smooth, how it does not want to mill anymore?

The ideology generated by the discourse surrounding the *metate* puts it in a very different place than the *comal*. Compared with the *comal*, which is characterized by complying with the daily grind, dying and then becoming reborn every day, the stone is defined by its perennial nature. It is quite possible that some of the stones owned by many of the Ch'orti' women are more than a century old. It is difficult for a stone to break and such events are poorly received, because the *metate* is often an inherited object and it is always desirable to pass it down, thereby linking and entwining generations and turning it into a true source of memories: it was given by one's mother and left to one's daughter. It is precisely because of this that a *metate* can also invoke conflicts between sisters. Also, beyond linking generations, the *metate* signifies a nexus in connecting the present world with the one that follows our death. This is illustrated by Ch'orti' stories about women who are said to continue milling in their afterlife.

New instruments in Ch'orti' kitchens: Metal mills, stoves, and iron *comales*

In the 1990s, the same blueprint implemented by the agricultural developmentalists of the 1970s was applied to foster change in the kitchens of eastern Guatemala, both in the local culinary technology and in everyday diets. The reference to "kitchen technology" was an indication of attempts to end technological simplicity exemplified by the everyday use of the milling stone and the cultural centrality of the *comal*. Both were seen as symbols of "primitivism" as well as signs of poverty, vividly illustrated by a diet primarily based on the *tortilla*. As developers often said, the milling stones were forms of Neolithic technology and the *comal,* arranged over three stones in the center of the home, was indicative of backwardness, lack of hygiene, and culinary uncouthness. The apostles of change also spoke of the advantages for indigenous peoples that would arise from the substitution of these artifacts. By banishing the *metate* and the *comal,* they said, the time a woman dedicated to the kitchen would be drastically reduced, and this would in turn reduce the excessive use of energy in the form of firewood (the fuel required for traditional cooking). However, behind the questioning of the *comal* and the *metate* there was a veiled attack against the cultural value of the *tortilla*. Indeed, advocates of economic formalism hoped that the new, modern tools would displace the potent ideological connotations of the corn *tortilla*. Within this logic, some nongovernmental organizations (NGOs) have been trying to introduce *nixtamal* manual mills made of metal, iron *comales,* and—in their most novel gambit—solar kitchens.

Mechanical metallic corn mills were first seen in local communities in the 1990s. Their importance as forces of modernization can be seen in the fact that they are used by political leaders as their gifts of choice during political campaigns. One candidate in the 2007 electoral campaign said during one of his rallies:

> Right now you can all come up to collect these small gifts we are giving you for the progress of [your] communities . . . You are all going to receive a *piocha* [hoe] so that you, farmers, can help yourselves in your daily chores . . . And all the women who show up are going to receive a corn mill; they will be able to take these mills and put them in their homes. You are not going to wear out your hands on the milling stone anymore; with these mills that we're giving to all of you, you can now mill your dough with ease . . .[5]

The NGOs seem to be just as eager to bring modernity to Guatemala's most forgotten regions. The leaders of the Zacapa Chiquimula Project

(FIDA-PROZACHI) highlighted with satisfaction how they had installed "228 manual *nixtamal* mills . . . given that the daily milling of corn for *tortillas* is a tiresome activity that takes a long time." For Ch'orti' women, even to this day, there are many things that make the metal mill inconvenient when compared to the *metate*. First, there is an element related to the object's aesthetic. They suggest that this artifact has displaced one of the most beautiful movements (according to local opinion) that the women make: the movement of the hips while milling with the *metate*. They find this movement pleasant and suggestive of bodily harmony. It can even be said that it has been a "traditional" expression of female courtship; by contrast, the movement of just one arm to operate the metal mill is neither creative nor beautiful for the Ch'orti'. Second, they are critical of its dysfunctional form: the mill's teeth do not allow the dough to become sufficiently smooth, with negative consequences in making good *tortillas* and, by extension, establishing "good" social bonds.

The centrality of the *comal* (and in an integral way, the ring of three stones it sits on) is also being attacked. Culinary modernity presents itself in many ways to the Ch'orti', who respond with mixed feelings to the presence of new tools. In 2012, we met Agustín, a young Ch'orti' peasant, at a convenience store in the municipality of Jocotán. He had just returned from the United States, where he spent five years working hard. He had built himself a cement house and wanted to furnish it in a manner befitting his new status. He was interested in a gas-fired iron *comal* with specifications that were written down on a sheet off to the side ("24" × 36" comal; 3 interior U-type burners [used] to produce propane gas-powered heat; Brick interior for heat retention; Double-reinforced tubular structure with a zincromate primer and a polychromatic Fast-Dry synthetic paint finish"). The shopkeeper explained him how good the item was, praised its durability, and highlighted its hygiene. Agustín did not buy it and, as he told us, he would never buy something like that for his wife: everybody in the community would laugh at her and he was sure that that device would not make good *tortillas*. The modernity he wanted to be part of (or wished his wife would join) did not stretch to owning that device.

However, if there is a clear example of the emotional gap between the traditional technology embodied by the *comal* and the new furnishings, it is the solar stoves. In 1996 we attended the launch of a project that intended to integrate this appliance into Ch'orti' culinary practices. The project was launched in the village of Suchiquer (Jocotán, Chiquimula) and all the promoters were there to introduce it. With the community gathered in one place, they described the four advantages of this stove: First, cooking was cheaper, because the stove did not require oil or heat to run; second, it was so easy to use that anyone could leave dishes to putter away and get on with other chores at the same time; third, it was healthier because the low cooking temperature preserved vitamins and minerals in the meal; and fourth, it was, in addition, good for the environment because it eliminated the need for wood and prevented deforestation. All this was asserted with the typical rhetoric characteristic of the verticality of development projects:

- "Do you want to stop the women in the community becoming ill from so much fire-building?"
- "Would you all like to return to the forests with the deer and the *coches de monte* (wild hogs) like there were back in the time of you grandfathers?"
- "Do you want to save a few bucks to help raise and educate your children rather than spend it on firewood?'

They also handed out a pamphlet in the form of a story titled *Solar Stove* that told the successful experience of this machine in the rural home of Doña Juana. She recounted what happened a few years back when she sent her son out to look for firewood. The child took a long time and finally returned with only a few small trunks: "Mom, I went all the way to the river and the only firewood there was all wet." Juana, with a worried expression and amid a landscape of devastating deforestation, reflected, "firewood is so difficult . . . you can't find it in the forest . . . it's so expensive already . . ." At this juncture, Aunt Pancha arrives, declaring that she cooks with the sun: "I realized that I could use the sun instead of firewood." Skeptical, Juana goes to Pancha's house to see how she cooks in a glass box. After the visit, and after eating a lunch that had been cooked in the solar stove, Juana says: "Oh Panchita, this food is so delicious. Where can we buy an oven like this?" The story ends with lessons on how to build the stove as a community, as well as some recommendations: start cooking in good time, because the solar cooking process takes longer; be sure that the glass panes are clean; place the oven in an open space, far away from trees and other objects that may cover the sun's rays; keep moving the oven every once in a while so that the reflector is facing the sun at all times; allow the oven to warm up for about an hour; leave the cover open after cooking until it cools down so that the humidity and steam inside can escape; maintain the oven well and store it away from rainwater; repaint it every so often and replace the reflective surface (aluminum foil) when it has lost its shine; never put hot pans on top of the glass because it could break; and avoid spilling liquid from the pans onto the metallic laminate in the back of the oven. A cooking chart accompanied the story listing the required cooking time for different products: steeped beans (five to seven hours); whole potatoes (three hours); chicken (four to five hours); meat (four to five hours); and vegetables (two to three hours). A diagram explained the oven's component parts.

The community accepted the project with alacrity. In barely three weeks, 250 homes had their solar stove, yet less than a month later, none of them was still in use. We have previously used the phrase "the ruins of projects of cooperation" (López García et al. 2011) making reference to the abandonment of infrastructures, devices, and services generated by cooperation projects. We highlighted how striking it was to encounter ruins of something that was built four or five years before. But the solar stoves surpassed even that: in less than a month they had been left out on *patios*, reduced to junk.

This case powerfully supports our argument that we need to consider the dimension of cultural affectivity involved in the use and adoption of culinary technology before undertaking changes. The solar stoves proved to have various disadvantages and to be incompatible with central aspects of Ch'orti' culture. In the face of the aesthetically pleasing and symbolically significant reality of the kitchen anchored in the home, here was a stove that had to be moved multiple times over the course of a day and "parked at night"—in the words of a common joke that quickly caught on. One of the project promoters later admitted to us that a woman from the community said between laughs that the solar oven "seemed like a *chivito* (goat kid) being moved around from one place to another." In addition, it was an uncomfortable and ugly device that, in her own words, required bothersome and costly maintenance. And most importantly, it cooked the grains poorly, leaving them too tough to achieve the smooth, soft dough needed to make a *tortilla galana*. One of the other peasants that took part in the project declared with less sarcasm: "If it can't cook corn, then why the fuck would we want a solar oven?!"

We are not suggesting that acceptance of technological change in Ch'orti' kitchens is impossible. We do, however, want to stress the argument that we have been advancing for some years regarding the need to consider the cultural affectivity in all processes of culinary change—from new dishes to new ingredients and new culinary technologies. Seeking to counter the impression that the Ch'orti' are resistant to change, we will end by briefly describing a project to modify kitchen technology which "has stuck" (*ha pegado*).[6] We are referring to the gradual displacement of the *poyetones*, a form of cooking that could be construed as an intermediate technology between the three-stone *comal* and the *lorena* stoves, with which the *poyetón* has a resemblance. The advantage of *lorena* stoves, compared with the *poyetón*, is their greater energy yield; still, their clear disadvantage is that they constitute a fire hazard in the many Ch'orti' homes with palm leaf roofs. The *poyetón* is a compact kitchen made from adobe with three openings: one for feeding the fire, and two or three burners that include a large one for the *comal* and two smaller ones for the pots used for cooking beans and making coffee. The problem is that although the *poyetón* distracts attention from the *comal*, which continues to be central, it still requires larger amounts of firewood than the *lorena*. Moreover, the *poyetón* does not have a flue, and thus smoke is not removed from the house, resulting in uncomfortable eye-watering and coughing for inhabitants. The *lorena* stove seeks to solve these problems.

The Guatemalan kitchen improvement program was started in the 1980s and aimed to reduce fuel consumption and improve the home environment. It dealt with a kitchen appliance made with large quantities of clay and sand, which results in the name *lorena* (a portmanteau of the Spanish words *lodo* and *arena,* for mud and sand, respectively), based on the local design for the *poyetón*. Beginning in the high plateau, where they were first built and where they are well established in the kitchens, the lorenas moved east with patchy results. In this region it was also PROZACHI that took charge of promoting them. Between 1993 and 1996, 3,336 stoves were installed in the region, resulting in energy cost savings estimated to be about 18,315 logs per family, and a series of benefits that, according to NGO's report, could be summed up as follows: a reduction in respiratory infections caused by smoke; little risk of burns in children; clean kitchens; and the ability to cook several things at the same time. But there is also an enormous attendant problem that has slowed down their adoption: they frequently lead to fires.

Thus, we can conclude that the problem of change in culinary technology is not to be found in the community's "tradition." Once again, the problem exists in the form of neocolonialism found among the promoters of change, even when it masquerades as paternalism. New devices are brought into local cultures with strong and consistent tales regarding their efficiency, beauty, and the advantages of certain foods and devices. The Ch'orti' people's resistance, far being the expression of blind "primitivism," constitutes the recognition of the cultural value of food. It may be that in the future we will find that they have ended up adopting iron *comales*, which last a long time but give *tortillas* a different flavor. It could also be that they will have banished the *metate* in favor of the metallic mills, even though they do not grind the corn dough finely enough, but maybe these adoptions are simply telling us that the Ch'orti' are moving on to another type of society. As the anthropologist Mary Douglas said, "The basic choices one makes are not between types of objects but, rather, between types of societies" (1998: 122), and to this day the Ch'orti' are not about to knock the corn *tortilla* off its nutritional and symbolic pedestal.

Notes

1 Although sorghum was first introduced in Guatemala in 1957 and named "maicillo," in fact, the Ch'orti' knew the existence of both true "wild" *maicillo*

and the farmed version, which was used as animal fodder and as an ingredient in the dish *delicados* (Wisdom 1940: 53).

2 The Feast of Spirits is celebrated in November, when it is believed that the souls of family members return to eat the food they are offered.

3 Doña Gregoria is referring to the unfortunate death of her five children over the course of a few months during a smallpox epidemic.

4 These four women (her daughters-in-law, her daughter, and her granddaughter) have had children in her house and had to eat *totopostes* (hard, toasted *tortillas*).

5 This kind of speech seeks to evoke the biblical curse, one to which the Ch'orti' have not paid much attention. In the local versions of the story of Adam and Eve, after having sinned, God tells them as punishment: "your fingers will become worn down on the milling stone and you will burn the palm of your hand on the *comal*" (López García 2003: 214).

6 This is very common indigenous expression for successful development projects, although it is also used for other businesses or even for relationships.

References

Douglas, M. (1988), *Estilos de pensar*. Barcelona: Gedisa.
López García, J. (2000), "La *tortilla* de maíz en el oriente de Guatemala. Estética y orden moral." Pp. 363–381 in *Anuario de Estudios Indígenas VIII, Instituto de Estudios Indígenas*. Tuxtla Gutiérrez: Universidad Autónoma de Chiapas.
López García, J. (2003), *Símbolos en la comida indígena guatemalteca. Una etnografía de la culinaria maya-ch'orti*. Quito: Abya-Yala.
López García, J., Arriola, C., Francesh, A., Nufio, E. and Mariano Juárez, L. (2011), *Valoraciones locales/retos globales de la cooperación. Un estudio de caso en Guatemala*. Madrid: Fundación Carolina.
Mariano Juárez, L. (2013), "El hambre en los espacios de la cultura. Visiones Indígenas Maya-Ch'orti'." *Revista de Antropología Iberoamericana* 8(2): 209–232.
PROZACHI, *Guatemala: PROZACHI: Proyecto de Desarrollo Agrícola para Pequeños Productores en Zacapa y Chiquimula (PROZACHI), 1991–1999*. http://www.ifad.org/evaluation/public_html/eksyst/doc/prj/region/pl/guatemala/r251gmbs.htm.
Wisdom, C. (1940), *The Chorti Indians of Guatemala*. Chicago: University of Chicago Press.

3

From bitter root to flat bread:

Technology, food, and culinary transformations of cassava in the Venezuelan Amazon

Hortensia Caballero-Arias, Instituto Venezolano de Investigaciones Científicas

Introduction: The centrality of cassava in the Amazonian diet

Despite its cultivation and consumption, cassava (*Manihot esculenta* Crantz) has commonly been regarded as a marginal staple within the fashionable gastronomic world and in the wider food industry. However, according to the Food and Agricultural Organization, cassava is the third most relevant source of calories in the tropical areas of Africa, Asia, and Latin America behind rice and maize. Although originally domesticated by native Amazonians, cassava was introduced into Africa in the sixteenth century by European explorers, often supplanting the cultivation of native African crops such as yam (*Dioscorea* spp.) (FAO 2009). By 2008, cassava world production was estimated at 233,391 tons, providing food to more than half a million people. Even though cassava is widely common and cultivated throughout these tropical areas, it is often perceived as a minor cultigen in comparison to other temperate crops.

Since cassava plays such a significant role in tropical agriculture, it is considered an important foodstuff that provides the primary source of carbohydrates in many areas of the tropics. In developing countries, this crop has been fundamental to both food security and income generation among small-scale farmers and indigenous groups. Thus, planting, harvesting, processing, and cooking cassava is a cultural lifestyle around food production and consumption.

A significant list of edible foods are byproducts of cassava processing, and these in turn are shaped by ethnic and national traditions. This crop is classified as either sweet or bitter, depending on its cyanogenic glucoside content. For it to be suitable for eating, bitter manioc requires a gradual and proper detoxification process. The bitter variety of cassava contains high levels of cyanide (prussic acid), and precisely this potential toxicity has been considered as one of the major limiting factors in its use for culinary purposes (Okezei and Kosikowski 1982). This is not the case for lowland indigenous peoples: indeed among many of Amazonian indigenous peoples, the bitter variety of cassava has been and still is a dietary staple, which is explained by a long-established technological processing system dating to pre-Hispanic times (Lathrap 1970).

My aim in this chapter is to explain, from an analysis of material culture, the techniques and technologies used in processing cassava among semi-urban indigenous communities located close to Puerto Ayacucho, the capital of the Amazonas State, Venezuela. In particular, my goal is to describe how these groups continue using traditional technology to make a cassava-based dish, commonly known as *casabe* or *cazabe* in Venezuela, considered as the "bread" of Amazonian peoples. Moreover, taking into account the centrality of this staple in Amazonian diet, in this chapter I address the difference between tradition and modernity as it relates to changes and adaptations of modern technology employed in the preparation of *casabe* destined for mass consumption in Venezuela. Above all, I seek to demonstrate that despite the incorporation of new technologies to increase *casabe* production for commercial purposes in urban settings, the traditional technological system of indigenous *casabe* has been instrumental in maintaining domestic relations of production among rural manioc smallholders. Thus, the preparation of *casabe* results from a direct association between an all-embracing technological system with an indigenous background and a food system based on a family and communal organization.

Technology and consumption: A historical overview of the cassava

Cassava is a tuber from the Euphorbiaceae family, and was originally domesticated in lowland areas of northern South America. Some authors

(Olsen and Schaal 1999) estimate that wild subspecies of manioc were early domesticated between 8,000 and 10,000 years BP in west-central Brazil. The origins, domestication, cultivation, and processing of manioc in the Amazon basin have attracted the attention of anthropologists and archeologists for many years. More recently, some studies among Amazonian peoples have focused on the nutritional implications of bitter cassava use, particularly the process applied to remove the toxicity of bitter manioc (Dufour 1995), the biological and cultural history of manioc, as well as the diversity of folk varieties cultivated (Rival and McKey 2008; Heckler and Zent 2008). Other works have emphasized the relationship between environmental and social factors with food intake variability and the diversity of bitter manioc cultivation systems in different Amazonian landscapes (Adam et al. 2009; Fraser 2010), as well as the symbolic and the intersubjective dimension of cassava (Mentore 2012). While these studies refer to cassava cultivation, varietal diversity and symbolism, the works of Carneiro (2000) and Westby (2002) have drawn particular attention on technological aspects of cassava processing.

Cassava or manioc, also known as *yuca* (in Spanish) and *mandioca* (in Portuguese), is a tropical annual crop that has adapted well to the less fertile and highly acidic soils of Amazonian lands.[1] It resists both long drought cycles and extensive rainy periods, as long as the roots do not become waterlogged (Schwerin 1970). Some Amazonian peoples believe that the best climate conditions for its cultivation are found in consistently warm weather with a relatively high level of humidity. That is why Lathrap (1970), taking into account the carrying capacity of the soils and environmental factors in the Amazon basin, stated that manioc is "one of the most productive and least demanding crops ever developed by man" (1970: 44).

As to the characteristics of these varieties, bitter manioc has a high starch content, it requires detoxification for eating and is more resistant to environmental stress; in contrast, the sweet variety yields less starch, it does not need detoxification, can be consumed boiled or roasted, and requires more fertile soils. Still, regardless of the variety, sweet or bitter, all cultivars contain cyanogenic glucosides; however, the latter is more toxic and needs additional processing before it can be eaten.[2] Because bitter cassava requires more time and energy to detoxify the tuber, one might assume that the sweet variety is preferred in these Amazonian settings. Yet a review of the distinct culinary functions of these two varieties reveals that sweet manioc should be used within a few days after harvest, while the bitter variety, once it is processed and reduced either as unleavened bread (*casabe*) or as flour (consisting of small globular pellets—*mañoco* or *farinha*), can be stored for some time before being consumed. Of all the plant foods in this tropical region, manioc produces more food per unit weight and is easier to carry (Wagner 1991).

It is important to note that there is a clear distinction between cassava[3] (referring to the plant and the root crop) and the "flat bread" called *casabe*, a derived meal. Due to the phonetic similarity, it is believed that the generic name cassava comes from *casabe* or *cazabe*, a Taino (Arawak) word that means cassava bread (Carrizales 1984). Thus, *casabe* is the flat round and thin bread made of bitter cassava flour, reported by Columbus on his first voyage to America in 1492 and described as "the roots [with which Indians] make their bread" (Colón 1985: 105). Today, *casabe* remains as one of the main dishes and source of dietary energy for most Amazonian indigenous peoples. It is also commonly consumed in the central and eastern regions of Venezuela, as well as in some areas of Colombia and the Antilles.

However, behind the *casabe* dish lies an ingenious process developed by Amazonian peoples to remove the hydrocyanic acid (HCN) from the tuber, which combines several techniques such as peeling, grating, squeezing, leaching, cooking, and drying, as well as technological artifacts such as knives, graters, squeezers (*sebucán*), sieves, griddles (*budare*), and baskets, among others. In fact, the evolution of cassava detoxification by using specific devices to squeeze manioc such as the *tipití* or *sebucán* has been the core of several debates and conjectures among anthropologists (Dufour 1995; Carneiro 2000). In this chapter, the purpose of a contemporary revision of cassava-based dishes is not to examine the origins of the artifacts but to emphasize that the consumption of *casabe* as edible food has implied an all-embracing technological system of indigenous tradition that has consistently maintained its association with the cassava culinary system.

Considering that technology, from an archeological approach, is the stage that comes between objects and societies (Dietler and Herbich 1998), and is also defined as any action that embraces "at least some physical intervention which leads to a real transformation of matter" (Lemonnier 1992: 5), I understand that technology is not only the material manifestation of cultural activity but also a social expression thereof. In this sense, Lemonnier (1992) suggests that "an anthropology of technology" must transcend the materiality of technological inventories and the study of the effects of technology on society. It should also deal with the relationship between technological systems and other social phenomena. From this point of view, both simple and more developed techniques respond to complex mental processes involving raw materials, energy, objects, gestures, and specific knowledge. The interaction of these components, the interrelation of technologies in a given society, and the relation between technologies and other social phenomena, make up what Lemonnier (1992) called a "technological system."

This phrase is useful as a contextual frame for *casabe* production among Amazonian indigenous communities. The relationship between technological systems and food systems seems like an obvious research area in anthropology nowadays, but there are few works that establish this inter-systemic connection. This is the case, for example, of an archeological study that reconstructs the ceramic wares used in food systems in the Venezuelan Andes through this approach (González 2011). At the same time, researches on food history and culinary practices in Venezuela have reached similar ideas about the complexity of food systems. Cartay (2005), for instance, asserts that any system or culinary regime has a compound structure based on a group of elements (including artifacts, ingredients, energy, and people), the relationship of these elements, and the operating rules in food selection, processing, and consumption. Therefore, the articulation of these technological and culinary systems will allow us to see that making *casabe* implies a series of actions involving raw materials, techniques, knowledge, and social actors, all of which interconnect. I will analyze the technological system of *casabe* by looking at the changes and continuities in its component elements among semi-urban indigenous communities.

Making *casabe* among indigenous peoples in the Venezuelan Amazon

Some years ago, while I was walking along the Orinoco Avenue in Puerto Ayacucho, the capital of Amazonas State, I saw that the stalls where indigenous families used to sell *mañoco* and *casabe* on the streets had been replaced by other traders, and above all by other—mostly Chinese-made—manufactured goods such as flashlights, batteries, pans, scissors, plastic utensils, and so on. Initially, I did not pay attention to this shift in the distribution of downtown urban commercial spaces, and assumed that the indigenous food vendors had been relocated to other places according to urban regulations. However, an indigenous Yanomami told me of his concern about how difficult it was to get cassava products in Puerto Ayacucho. He was worried because of the extended shortage of these Amazonian foods. He said: "Ayacucho used to be full of *mañoco* and *casabe*. What happened to all of those staples? Why do the Indians no longer produce manioc?"

This particular comment on the scarcity of *casabe* and *mañoco* in such an important location made me wonder about the processing and consumption of these products in an urban Amazonian setting. It was reasonable to ask whether peripheral indigenous communities of Puerto Ayacucho still used the same traditional technology and cooking practices to make *casabe*, or if new objects (tools, instruments) had been recently incorporated. I also wondered whether the lack of these products at that time could be attributed to other causes, such as the impact of climate change on manioc harvesting or the existence of radical cultural transformations linked to new public policies.

As I discovered, this latter hunch proved correct. Between 2007 and 2010, many indigenous communities were the recipients of several social programs (social missions) organized by Venezuelan state institutions that granted credits along with funding for construction and access to manufactured goods that diverted their agricultural practices. During that time, the metropolitan indigenous communities of Amazonas State were more concerned with complying with these aid policies than they were with maintaining their gardens and harvesting cassava or any other crops. To make up for the cassava shortfall, the two neighboring states of Bolívar and Apure provided *casabe* and *mañoco* for local consumption in Puerto Ayacucho, and even communities that were traditionally recognized for their high-quality production, such as San Juan de Manapiare, had to procure these food staples from other regions of the country. This situation developed some time later when the elders of various villages complained that the shortage of these staples was affecting their internal subsistence dynamics. About four years ago, indigenous communities in these peri-urban areas returned to their gardens to cultivate manioc and resumed making *casabe* and *mañoco*. "For a while we stopped producing cassava, but now we have rediscovered what we've known how to do for centuries," said one of the producers from a Piaroa (Wotjuja) community.

Currently, *casabe* is produced by most of the indigenous peoples in the Venezuelan Amazon. It is regularly made by indigenous peoples of Carib and Arawak descent, as well as other linguistically independent ethnic groups, such as Piaroa and Yanomami, who inhabit riverine and inter-fluvial zones in this tropical rainforest region. These traditional communities harvest and prepare cassava as a staple food. However, in the case of peri-urban indigenous communities, they have experienced significant changes that have affected, among other things, the preparation of *casabe*; based on these cultural shifts, it seems relevant to account for the current transformations and continuities in *casabe* production and its related technologies.

The ethnographic data for this work was obtained during short-term case studies conducted in three culturally different indigenous communities belonging to Kurripako, Yekuana, and Piaroa (Wotjuja) peoples. Kurripako is an Arawak-speaking group, Yekuana is Carib, and Piaroa is a linguistically independent people. All three communities are located no further than a half-hour drive from Puerto Ayacucho and are settled in savanna environments characterized by acidic and dry soils. Territorially they are part of the Atures municipality of the Amazonas State. This municipality has the highest population concentration of this territory. According to the 2011 census, the population of Atures was estimated to be 104,228, of which 36,004 were indigenes. This means that Puerto Ayacucho and its peripheral areas contain a significant number of indigenous communities that constantly experience transcultural as well as intercultural processes. Some of them used to live in rainforest landscapes in the fluvial Orinoco–Río Negro axis but for a variety of reasons they have migrated to areas near Puerto Ayacucho over recent decades. Despite this mobilization from their places of origin and the urban expansion toward their communities, they endeavor to reproduce their local ways of life by keeping up their agricultural practices; one of them is manioc cultivation. The unit of analysis was the household, which comprised extended families in all three cases.

Among Amazonian indigenous peoples, cassava appears mainly in two different forms: as *casabe* (bread) and *mañoco* (toasted flour). Depending on the ethnic origin of the maker, one person's *casabe* can be thicker, thinner, sourer, softer, or more toasted than another's, but overall it is the common carbohydrate that complements such protein intake as fish or game. *Mañoco*, on the other hand, is commonly consumed with water, in a beverage called *yukuta*, or used to thicken soups. Regardless of the differences in flavor and texture of *casabe*, the overall operational chain in processing manioc is quite similar among indigenous communities. However, what varies is the incorporation of new technological tools described below.

Making *casabe* implies establishing a distinctive operational sequence. Among the Tukanoans of northwest Amazon, Dufour (1985), for instance, examining the relationship between time and energy consumption, describes three phases for the daily manioc routine: acquisition, processing, and cooking. Rehearsing some steps of this scheme, a complete description of the processing and cooking stages of manioc in the Venezuelan Amazon goes as follows: Once the manioc is harvested, the roots are peeled. They are then washed well before being grated, the roots reduced to a watery mash. The grated pulp is collected in a container, and then stored and fermented for a day or two. The fermented mash is subsequently placed in a *sebucán* (a long cylindrical woven basket plaited from the strips of the *tirite* plant [*Ischnosiphon arouma*] and with a loop at each end). Once filled, the *sebucán* is hung by the upper loop and subsequently pulled by adjusting the wooden lever that passes through the bottom loop. The lever is brought down to fit into the notches of a pole fixed to the floor in a sequence of moves. The *sebucán* is pulled, squeezing the pulp and yielding the manioc juice called *yare*. This liquid is collected in a pot and can be later used in the preparation of other dishes. The pulp, which is partially dry and relatively free of prussic acid, is then removed from the *sebucán* and passed through a woven basket (*manare*) to sift out the thin flour (*catebía*), which indigenous peoples use to produce either *casabe* or *mañoco*. To make *casabe*, the flour is spread out as a round cake on a preheated griddle. The heat compacts and roasts the flour forming the *casabe* cake. Once it is cooked on both sides, it is left out in the sun for a few hours to eliminate any remaining humidity.[4] To make *mañoco*, the flour is spread on a griddle and tossed until it is dry; it is now an easily-transported coarse meal.

Indigenous communities of the Venezuelan Amazon follow this operational sequence for making *casabe* almost without exception. In this technological system as a whole, however, there are some elements or components that are more difficult to change than others. By correlating the processing and cooking phases of cassava with the cogs of the technological system, I examine ethnographically to what extent semi-urban indigenous communities have incorporated new technologies into their traditional patterns of cooking *casabe*, and whether that change has modified the perception and consumption of this product within modern conceptions of making *casabe*.

In the following paragraphs I examine in further detail the different steps involved in the process of *casabe*-making: The first step consists of peeling the manioc. It requires tubers as the main raw material, as well as tools such as knives, containers with water in which the peeled cassava can be stored, places to sit, and enough outdoor space to be comfortable during this endeavor. This arrangement will also allow the peelers to pile up the manioc rinds nearby. During this activity, the energy and gestures come from a repetitive manual dexterity in knife use. Usually, an individual grabs one tuber, scrapes it six or eight times, and then throws it into a plastic bucket with water, while setting the rinds aside. This procedure requires fairly mechanical knife skills. All the women from the household participate in this activity and in some cases children and young males help them peel large quantities of manioc. At this stage, it is clear that the amount of cassava peeled will be the portion used for the final production of *casabe* cakes. In all of the three communities we studied, they wash and later leave the peeled cassava in water for several hours or even a whole day—the first step in the fermentation process. The soaking time will depend not only on the ethnic tradition but also on family customs replicated from generation to generation.

The second step is grating the peeled cassava. Depending on the type of grater available, more energy and more people may be required to work their way through the tubers. Manioc graters have changed considerably over time: according to archaeological findings they have evolved from stone to wooden implements. For example, Carrizales (1984) reports that one of the Taino techniques for grating cassava during pre-Hispanic times involved the use of stone graters covered with sharkskin. Among Amazonian indigenous peoples in Venezuela, wooden grating boards set with stone chips or nails are still used. This type of traditional wood grater was found among the three indigenous communities where I carried out fieldwork, but women said that they used it only when they did not have a grating machine. In fact, they did not use it in any of the three communities while I was there. In turn, the mechanical grater, called *cigueña*, which functions with gasoline and oil, has become increasingly important in the process of making *casabe* during the last twenty years. Due to this technological shift, indigenous communities have been able to reduce time- and energy-consuming work, while learning how to use a machine that requires skill and care to place the cassava correctly. By changing the type of object—the grater—gestures and uses of spaces are now different. When using the wooden grating board, women had to kneel or stand up with their backs bent while moving the arms up and down to grate the cassava. Today, however, they can avoid this uncomfortable, energy-sapping process by using a machine that fits in with their environment and indigenous domestic dynamic. They usually purchase such devices from a specialist shop in Puerto Ayacucho for the equivalent to 430 *casabe* cakes as they are sold in the city's principal market on Saturdays. Given its high cost, it is no wonder that such machines are prized objects in households. After grating the cassava, the producers place its pulp in a wooden container and leave it outdoors for a day or two to ferment. However, the time of fermentation may vary depending on the producers' ethnic background and on household recipes.

The third step, squeezing the cassava pulp, is perhaps the most striking technique because of the use of elaborate objects such as the *sebucán* and the sophisticated wooden lever piece. The material complexity of these artifacts involves the utilization of the following resources: the natural fiber *tirite* (*Ischnosiphon arouma*), wooden sticks, and the cassava pulp. It takes energy and knowledge to plait the reed strips in the right way to assemble the *sebucán* and to adequately place the wooden lever and pole. During this process, the woman in charge of squeezing the juice or *yare* has to perform several tasks: she has to fill the *sebucán* with grated manioc pulp, then press down the wooden lever, all the while avoiding the poisonous juice dripping from the *sebucán*. The woman overseeing the process will also have to sit on the end of the lateral stick to gradually adjust it and to increase the pressure on the *sebucán*. She presses down on the lever with all her weight so that it can reach the lower notches of a pole. These actions demand a great deal of energy as well as specialist knowledge of instruments that themselves require a sequence of bodily gestures and language that must be repeated many times in order to successfully detoxify the cassava. In two of the communities I visited I found *sebucanes* made out of plastic straps. One of them was from Colombia, and according to the owner, the *sebucán* had been exchanged for other goods. The structure of this plastic artifact is quite similar to the one made out of reed strips, and squeezes the pulp well. In fact, one of the women indicated that both two types of *sebucán* were effective, but the plastic one lasts much longer.

The final step of processing cassava prior to cooking is sieving the grated manioc mash. At this point the raw material consists of compacted cassava pulp; circular baskets are used to sieve it before it is stored in plastic receptacles. (All three communities in the field study used traditional baskets woven from *casupo*, the *tirite* plant [*Ischnosiphon arouma*]. Passing the thick flour through the sieve is an energy-intensive process for the women, but once sifted, the flour can be cooked. The sieving is a standard manual technique that any member of the household can perform; even the children sitting on the ground mimic the women's gestures when they help sifting the flour.

The next stage is to cook *casabe* itself, which requires the pulp, sifted flour, and items such as a hearth filled with plenty of wood, the griddle (*budare*), a plain woven pallet, and a basket (*manare*) in which to place the final product. A special cooking technique is used to spread the flour on a hot griddle and form a flat cake of approximately thirty inches in diameter. Depending on taste preferences and ethnic tradition, the woman in charge of cooking will make the cake thick or thin, displaying skillful hand movements over the flour dough. She has to be dexterous and careful not to burn the *casabe* or herself, and she must also flip the compact cake at the right moment, all the while ensuring that the hearth burns consistently. Once the *casabe* is cooked on both sides, she places it on the *manare* basket before taking it outside to dry on top of a roof or a metal mesh specifically made for this purpose. Cooking is thus the final stage of a long operational process necessary to modify and process the raw material—that is, to detoxify and dry it out—so that it is rendered edible. Complex tools, time, and energy also play their part, as do a thorough understanding of the physical techniques and specialist knowledge required to produce the *casabe* cake.

When considering the operational sequence necessary to make *casabe*, it is important to note some general issues. First, it is striking that indigenous communities, and also non-indigenous peoples, use a common terminology to identify artifacts and substances such as *sebucán*, *manare*, and *yare*, regardless of ethnic origin. This shows that around *casabe* production there is a shared meaning concerning terms and expressions that transcends ethnic boundaries. Second, regarding gender roles in manioc production, indigenous women are always in charge of harvesting, processing and cooking cassava. A man can accompany a woman to the gardens, and then perform other tasks or take care of the children while she harvests the crops. He can also look for firewood to be used in the hearths when the flour is cooked, but he will not participate in the preparation of *casabe* itself. The skills, the knowledge of the quality of cassava as raw material, and the use of tools and instruments, are all shaped by the woman's age and abilities, which vary in terms of energy, time, and physical movements. There is thus a female indigenous habitus embodied in making *casabe* that is shaped by the materiality of the objects, among other cultural characteristics. Third, the kitchen spaces for making *casabe* are divided into outdoor and indoor places. The outdoor spaces are where cassava is peeled, grated, and squeezed. By contrast, the space inside a shed or hut is where the hearth and the griddle are located, and it is there that the *casabe* is sieved and cooked. This spatial pattern was similar in all three indigenous communities. The uses of multiple spaces in the operational chain reflect that instruments, energy, and knowledge are relevant in the technological system, but also important is the appropriation of different spaces according to the activity performed.

Conclusions

In relation to the extremely time-consuming activity that represents cassava processing, Dufour (1995), in the case of the Tukanoan, stated that as women's expectations could change with acculturation, cassava-processing techniques might also change. In this case, the ethnographic data has suggested that technological changes have helped cut down the time and energy that needs to be expended in this activity. Nevertheless, the incorporation of new technological instruments into the process of making *casabe* has not affected the social relations of production, the final product for consumption, or the perception of whole process in the three indigenous communities studied. However, the assimilation of larger machines and plastic *sebucanes* in processing manioc has constituted a significant alternative for indigenous women to channel the energy in a more advantageous way.

It is understandable that the use of these new technologies has influenced the utilization of energy, gestures, and technological know-how among the producers, but by reviewing the operational sequence as a whole I found that there is no tension between the traditional technological system of making *casabe* and the adaptation of new instruments. In sum, the data account for a correlation between the household dynamics and the technological system of *casabe* production that has been maintained over the years despite the cultural changes and the incorporation of new tools and instruments in these semi-urban indigenous communities.

Acknowledgments

This paper draws on ethnographic field trips to the Amazonas State funded by the Venezuelan Institute of Scientific Research (IVIC) between August 2014 and January 2015. I am thankful to Héctor Blanco for his insightful comments on the socio-cultural dynamics of Puerto Ayacucho and to Yheicar Bernal for his support and collaboration during fieldwork. This paper has benefited from the readings of Krisna Ruette, Erika Wagner, and Felipe Caballero Jr. I would also like to thank the editor, Steffan Igor Ayora-Díaz, for his invitation to contribute to this volume.

Notes

1 Fraser (2010) argues that there has been an oversimplification regarding how successfully bitter manioc has adapted to infertile soils, and demonstrates that this crop can yield well in alluvial and fertile soils of the Amazonian whitewater landscapes.
2 The bitter root has cyanide concentrations greater than 100 ppm, while the sweet one has less than 100 ppm (Dufour 1995).
3 "Cassava" is the common name of this crop in English, although some authors argue that the correct term is "manioc" (Gade 2002).
4 The photographic work of Thea Segall (1979) was useful to illustrate the step-by-step sequence of *casabe* preparation.

References

Carneiro, R. L. (2000), "The Evolution of the Tipití: A Study in the Process of Invention." Pp. 61–93 in G. M. Feinman and L. Manzanilla (eds), *Cultural Evolution: Contemporary Viewpoints*. New York: Kluwer Academic/Plenum Publishers.
Carrizales, V. (1984), "Evolución Histórica de la Tecnología del Cazabe." *Interciencia* 9(4):206–213.
Cartay, R. (2005), *Diccionario de Cocina Venezolana*. Caracas: Alfa Grupo Editorial.
Colón, C. (1985), *Diario de a bordo*. Madrid: Ediciones Generales Anaya.
Dietler, M. and Herbich, I. (1998), "Habitus, Techniques, Style." Pp. 232–263 in M. Stark (ed.), *The Archaeology of Social Boundaries*. Washington: Smithsonian Institution Press.
Dufour, D. L. (1985), "Manioc as a Dietary Staple: Implications for the Budgeting of Time and Energy in the Northwest Amazon." Pp. 1–20 in D. J. Cattle and K. H. Schwerin (eds), *Food Energy in Tropical Ecosystems*. New York: Gordon and Breach.
Dufour, D. L. (1995), "A Closer look at the Nutritional Implications of Bitter Cassava Use." Pp. 149–165 in L. Sponsel (ed.), *Indigenous Peoples and the Future of Amazonia*. Tucson: University of Arizona Press.
FAO (2009), Food Outlook. Global Market Analysis. http://www.fao.org/docrep/012/ak341e/ak341e06.htm [accessed 6 January 2015].
Fraser, J. (2010), "The Diversity of Bitter Manioc (*Manihot Esculenta* Crantz) Cultivation in a Whitewater Amazonian Landscape." *Diversity* 2(4):586–609.
Gade, D. W. (2002), "Names for *Manihot esculenta*: Geographical Variation and Lexical Clarification." *Journal of Latin American Geography* 1:55–74.

González J., N. (2011), *Mucuras, oscios y budares. Reconstrucción de equipos cerámicos desde la perspectiva de los sistemas tecnológicos*, Master's Thesis, Center of Anthropology, Venezuelan Institute for Scientific Research, IVIC.

Heckler, S. and Zent, S. (2008), "Piaroa Manioc Varietals: Hyperdiversity or Social Currency?" *Human Ecology* 36:679–697.

Lathrap, D. W. (1970), *The Upper Amazon*. New York: Thames and Hudson.

Lemonnier, P. (1992), *Elements for an Anthropology of Technology*. Ann Arbor: University of Michigan.

Mentore, L. (2012), "The Intersubjective Life of Cassava among the Waiwai." *Anthropology and Humanism* 37(2):146–155.

Okezei, B. O. and Kosikowski, F. (1982), "Cassava as Food." *Critical Reviews in Food Science and Nutrition* 17(3):259–275.

Olsen, K. M. and Schaal, B. A. (1999), "Evidence on the origin of cassava: Phylogeography of *Manihot esculenta*." *Proceedings of the National Academy of Science* 96:5586–5591.

Rival, L. and McKey, D. (2008), "Domestication and Diversity in Manioc (*Manihot esculenta* Crantz ssp. *esculenta*, Euphorbiaceae)." *Current Anthropology* 49(6):1119–1128.

Segall, T. (1979), *El Casabe. Secuencia fotográfica*. Caracas: Editorial Arte.

Schwerin, K. (1970), "Apuntes sobre la yuca y sus orígenes." *Boletín Informativo de Antropología* 7:23–27.

Wagner, E. (1991), *Más de quinientos años de legado americano al mundo*. Serie Medio Milenio Cuadernos Lagoven. Caracas: Editorial Arte.

Westby, A. (2002), "Cassava Utilization, Storage, and Small-scale Processing." Pp. 281–300 in R. J. Hillocks, J. M. Tresh, and A. Bellotti (eds), *Cassava: Biology, Production and Utilization*. Wallingford: CABI Publications.

4

Technologies and techniques in rural Oaxaca's Zapotec kitchens

Claudia Rocío Magaña González, Centro de Investigaciones en Comportamiento Alimentario y Nutrición (CICAN), Universidad de Guadalajara

Introduction: Kitchen space and technologies

Rural Mexican communities are constantly adopting, adapting, and sometimes rejecting modern culinary values. In this chapter, I explore the regional, local, and ethnic dimensions of rural kitchens in the southern Mexican state of Oaxaca, the local cuisine of a Zapotec community, and their transformations from one generation to another. I describe how two traditional cooks[1] from an indigenous community in the Isthmus of Tehuantepec combine modern and traditional values, techniques, and technologies, examining rural kitchens as spaces and regional cuisine as symbolic representations of their ethnic identity. During three years of fieldwork in Oaxaca, I found that Zapotec women are able to build, rethink, and negotiate their ethnicity through the process of blending and revalorizing ethnic food in everyday domestic practices and during religious feasts featuring iconic cultural meals.

In the Zapotec communities of the Isthmus of Tehuantepec, two types of rural kitchens exist: household and community kitchens. In both spaces, cooks combine traditional and modern techniques and technologies. Although household kitchens are generally permanent spaces, some have been modified in accordance with family needs while still preserving the basic structure required for cooking daily meals. On the other hand, during *fiestas*, traditional Zapotec cooks often incorporate new foods into regional cuisines. In these ritual spaces, community members build temporary kitchens at the *fiesta* site. These community kitchens emerge and disappear in different locations with each collective ceremony. Both types of rural kitchen illustrate how local tradition is anchored in ethnic identity and values. Over time, women integrate goods as well as symbols into their collective knowledge and beliefs, reflecting changes in the regional cuisine. As a form of cultural practice, cooking in these communities allows individuals to rethink and negotiate the meaning of belonging to an ethnic group in the twenty-first century.

Contextualizing *Istmeño* cuisine

The term "cuisine" can have different meanings, such as: (a) goods produced in a kitchen; (b) culturally differentiated products; and (c) specialized procedures of cooking that appear among some social groups (Goody 1995). I find the third meaning useful to analyze techniques and technologies found in the *Istmeño* cuisine within a specific physical and social region in the state of Oaxaca, Mexico. However, the people of the Isthmus of Tehuantepec tend to complement the practical and material dimensions (techniques and technologies) of a cuisine with affective and discursive narratives. For example, Henestrosa, one of the most popular acculturated Zapotec indigenous writers of the twentieth century, argued in his writings that because of its blend of indigenous and Spanish creativity and ingredients, Oaxacan cuisine was the largest and the richest in Mexico. The idea of *mestizaje*[2] repeatedly appears in his literary work,[3] particularly when describing the intelligence and sagacity of Zapotec women who knew how to blend and mix Spanish ingredients such as olives, grape leaves, and grapes with their locally produced counterparts, like chili, achiote, beans, squashes, and *chayotes* (Henestrosa 2000). In his opinion, this explains why *Istmeño* cuisine is such a significant part of Zapotec ethnicity in this part of Oaxaca.

The geography of the region helps explain the importance of cuisine in the Zapotec culture. The Isthmus of Tehuantepec is one of seven regions of the state of Oaxaca in which various ethnic groups coexist near the Pacific Ocean. There are mountains on the east side of the region, and plains and coast on the west side. To the north lies a valley surrounded by the Sierra Atravesada and the mountains of Ixtaltepec; to the south one finds the coastal belt of the Pacific Ocean. These different ecosystems host a great variety of flora and fauna used as ingredients, and these account for the richness of local cuisine.

Livestock production, agriculture, and fishing are the principal economic activities in the region. Thus, the ingredients that form the *Istmeño* cuisine can vary from the occasional inclusion of wild animals such as iguanas, wild pigs, and deer, to the more frequent use of domestic animals in the everyday diet—cows, chickens, and pigs, as well as different types of fish, crabs, and even turtle eggs.

Another important aspect of *Istmeño* cuisine is the cultural and social diversity of the area. The region encompasses six different ethnic groups (Zapotecs, Huaves, Mixes, Zoques, Chontales, and Choles), all of which maintain their cultural differences. Although the history of the coexistence of these ethnic groups is profoundly complex, their diverse diets represent the symbolic and practical sources of ethnicity[4] in the region. Thus, food production offers a setting in which we can understand how interethnic relationships are continuously established and negotiated in the region.

The Isthmus has a long history of power inequalities among ethnic groups (Magaña González 2012). The Huaves were the most powerful ethnic group until the 1300s, when the Zapotecs arrived and began to conquer the territory. During the following centuries, people on the Isthmus participated in several revolts related to struggles to control the riches of the region (Coronado Malagón 2000; Tutino 1980). Over time, and especially since the nineteenth century, Zapotec leaders have established strategic alliances with political and economic elites, setting themselves up as the region's dominant group. Historical, economic, and political dimensions converge in social and culinary arenas, where members of different ethnic groups continuously articulate their discourses and practices (Magaña González 2007, 2012). Throughout the region, social complexity is marked by interethnic power relationships that drive negotiations and agreements that, in effect, promote the culture of each group; in this case, through their cuisine.

For example, anthropologists and sociologists have recorded how different ethnic groups adopted and adapted some Zapotec practices during the historical formation of regional hegemony. This process has been called the "Zapotequization" of the region. However, this does not mean that the Huaves, Zoques, Mixes, Choles, and Chontales surrendered their cultural practices and symbols when they came under Zapotec domination. On the contrary, subordinate groups adapted and adopted symbols throughout an ongoing resistance of conquest, and unfolded them as part of the development of *Istmeño* regional identity (Coronado Malagón 2000; Magaña González 2012; Scott 1985). This identity category allows members of an ethnic group to affirm that their food, music, literature, and dress are regional symbols and not a Zapotec imposition. Therefore, the term *Istmeño* refers to the place (family and town) where people of the region were born and raised, as well as where they live and share cultural aspects that are a product of historical negotiations between members of different ethnic groups. *Istmeño* also refers to a shared way of life found in the region.

However, for the purpose of this analysis, I emphasize the importance of Zapotec dominance as it has been described and discussed in academic writings, as well as in the political speeches of the region's cultural leaders. Evidences of Zapotec expansion can be found in different arenas—such as the economy, politics, and culture—and has influenced Huave *fiestas* (Magaña González 2012). This expansion began during the nineteenth century, when the Zapotecs reclaimed their ethnic identity as part of a powerful discursive strategy to resist various modernization projects in the region, especially during the presidency of Porfirio Díaz (1876–1911). At a regional level, some Zapotecs gained advantageous political and economic positions as middlemen, spawning an intellectual and military elite (Millán 2005; Reina Aoyama 1997). During that period, Zapotec women, who during the nineteenth century marketed produce along the railroads, developed ideas and practices concerning regional food production (Coronado Malagón 2009). The role of Zapotec women remains important today, not only becuase they are visible in market places preparing and selling food, but also because historically they have been the spokeswomen of Zapotec traditions.

Rural household kitchens: From domestic cooking to the sale of everyday regional dishes

A traditional *Istmeño* household kitchen is located outside the house. This is due in part to cultural and climatic reasons. As in many cultures, in this location kitchens are places where food is prepared using wood fires and following specific culinary practices (Lévi-Strauss 1964, 1965). Wood fires heat the space, making the room very hot but also ensure that the correct flavor can be obtained. Since temperatures in the Isthmus are usually above 25 degrees Celsius, traditional rural kitchens have no walls, and consist of four wood poles that support a gabled thatched roof made of palm leaves. The cook reserves one side of the space for preparing food. In this area she places hot pots on the fire located on a stove (with no grill) made of concrete or adobe bricks. The upper bricks are movable, allowing the cook to arrange them depending on the size of the pot and other needs. At one side of the fireplace there is an aperture for the oven (*comizcal*). The oven consists of a pot of clay filled with bricks and sand, and the top part is left open. These days, the *comizcal* is rarely used: since it has many disadvantages when compared with a modern oven, the younger women tend to dislike cooking in them. To use a *comizcal* effectively, cooks need to develop technical abilities and acquire the necessary knowledge to light the fire inside the hole and to recognize when it is hot enough (when the fire is reduced to embers, without a flame) for *totopos* and *gendarmes*[5] to be placed on its the walls. They must constantly check that the food is not falling into the ashes, and also know when to take the *totopos* and gendarmes out the *comizcal* without burning themselves. Thus cooks need years of observation and practice before they can achieve a perfectly cooked and consistent *totopo*. Another important area of the kitchen is the sink, where the cook washes the dishes or the food. Most houses in the region lack indoor plumbing and hence cooks must use hoses to fill containers with water.

The kitchens also contain old racks with dishes and modern appliances such as blenders. One can also find *molcajetes* (a stone mortar used for chili peppers), *pichanchas* (a strainer made of *guaje*[6]), plastic dishes, large wooden and plastic spoons, and cutlery in this area. Old metal and aluminum pots can be found in random places: on top of a chair, on the table, or on the dirt floor. Each cook organizes her space according to her own logic of efficiency. Thus, "traditional" household kitchens may not look organized or well designed to an outsider, but for the cook they are functional.

In these kitchens one can also find old wooden tables where cooks grind ingredients, chop food, and chat. Tables are important because Zapotecs make many decisions, negotiate, and argue about both food and other household issues around them. Cooks on the Isthmus typically acquire new kitchen instruments for practical reasons, but may receive a can, a bottle, a soup ladle, or a pinafore dress as payment for her help in catering for a ritual. On the other hand, older cooks may possess an appliance such as a blender not because they wanted or needed one, but rather because it was given as a gift by a relative. Many older cooks realize that the younger generations find advantages in certain appliances that allow them to prepare food better and faster but in general local people feel that more mature women, especially the most renowned traditional cooks, achieve the desirable flavor and quality of food thanks to their use of traditional tools such as the *molcajete* (stone mortar) and the *metate* (stone mill).

In the Isthmus kitchens are the places where cooks prepare food for the family, and women also use these spaces to cook food that will support their families economically. It is common among Zapotec women to sell "home-made food" as a strategy to supplement household income, especially when their children are young. Typically, though, people may purchase food at these women's homes, but do not eat it there. A second form of food commercialization involves selling "door-to-door," which means that after the women have cooked at home, their sons or daughters use freight trikes (*triciclos*) to carry and sell plastic dishes, aluminum pots, or pottery containing the food along the dirt roads throughout the community. A third mode is "selling food at home" as a formal business, requiring space modifications and adaptations, the purchase of kitchen and service equipment, and a distribution of chores among family members (to buy products, to prepare, serve, and charge for the meals and cleaning). The first mode responds to the logic of subsistence and reproduction, while the second and third respond to "capitalist rationality" (Magaña González 2007).

As spaces, household kitchens on the Isthmus of Tehuantepec have experienced modifications over the past three or four decades, largely in response to migration processes (Magaña González 2007). For example, nowadays, traditional and modern kitchens can coexist within the same house: variations include new plumbing in traditional kitchens, as well as houses with no traditional kitchen but a new one inside the house equipped with a gas oven on which continue to use large pots. Very often, changes in kitchens are gradually influenced by younger generations, especially when they return to their village after studying or working in different cities in Mexico or in the United States. In such places, migrants discover decorations or instruments that can be adapted or integrated into traditional kitchens. Decisions regarding a new arrangement of the kitchen space rest with the cook, although older family members participate in the negotiation process. Nevertheless, cooks are generally open to modifications of their kitchens. First, they value the experience of sons and daughters returning home because they see them as having progressed in life as a result of their exposure to modern, urban societies. Second, when family members return, they bring the money that facilitates such "improvements;" and third, the Zapotec consider these adaptations a symbol of prestige[7] within the community. To understand how the communities sustain and transform technologies and techniques over time, I find it useful to highlight the relationship that Zapotec cooks establish with the traditional rural kitchen as a material space where they can display their knowledge and cooking skills. Thus, in the relationship between cooks and kitchens, it is possible to value and transform culinary practices, food production, cooking instruments, and technologies.

I met Na[8] Maria because I wanted to learn about traditional cooks in a small town located at the eastern part of the Isthmus of Tehuantepec. Outside the house where I had been told she lived, I shouted her name. Suddenly, a man holding a large knife and with bloodstains in his shirt appeared from behind the black metal door, and told me to come inside. I saw a rectanglular yard, of no more than 30 x 20 meters. At the end of the yard, there was a house built from concrete blocks. On the right side of the yard there was a large wall marking a private property, which was higher than the single-story house divided in three rooms where Maria's family lives. There were two different constructions visible on the left side. The first one was where Rutilio, the youngest of Maria's sons, worked as a butcher. His instruments—an old wooden table and some metal hooks—hung from a pipe that supported the metal roof. At the end of Rutilio's workplace was Maria's kitchen, the revered space where affect and custom are inscribed into everyday cooking practices.

When I arrived at Maria's home, she demanded to know who had told me about her. I responded that every time I asked about the traditional cooks in the community, people mentioned her and also that many people mentioned her as a cook who "sold food at home." The first interview lasted more than two hours, and led to over three months of repeated interaction. Before I knew it, we were folding banana leaves, filling them with the special mixture to make black *mole tamales* that Na Maria could sell at home, just as she had done every day for the past four decades. What must a traditional cook know, feel, and achieve during her life to be recognized by the community as an outstanding cook proficient in traditional culinary practices?

When we met, Na Maria told me that she had spent more than forty years cooking and selling food at her home. At that time (2006), she was eighty years old. She said that cooking was her life, and at the same time it was the only way that her body would continue working, because the doctor said that keeping active would keep her healthy. Na Maria told me that she had learned to cook from an older friend when she accompanied her to the *ayudas*[9] three or four times a week. Her friend was well known locally because she was the first woman to charge money for cooking in community rituals back in the 1960s. For such events, Na Maria learned how to cook traditional dishes such as black *mole*, *tamales*, *frito de res* (beef stew), and *botanas* (different tapa-like dishes served during feasts). Na Maria also learned different cooking techniques when she was little so that she could prepare food at home before her mother died. She would grind corn by hand to make *totopos* (hard *tortillas*) and put them into the *comizcal*.

When she started to accompany her friend, Na Maria learned specific details of recipes, such as the quantity of *mole* that had to be prepared for a wedding. Her learning technique was to pursue a regular practice in *fiestas* and at her home: "trial and error." For example, when I asked her how a traditional dish was cooked, she explained technical procedures in a random order, because she does not find them to be the most important part of cooking. For instance, the *barbacoa* needs to be cooked after being seasoned with special spices in special aluminum containers. Banana leaves are placed on the bottom of the casserole but must be dried so that they can be used simultaneously as covering and seasoning. Subsequently the meat is placed in the casserole. The commensal's opinion, and the cook's experience, combine to result in a dish that can be considered as "well cooked." While Na Maria prepared many dishes for fiestas, it was at home where she developed the art of cooking large amounts of food.

When Maria was forty years old, she divorced her husband because she could not take any more physical and psychological abuse. The real reason why she started to "sell food at home" is that she needed money to gain economic independence. She decided that her culinary knowledge could help her do this while her children were small and needed her care. During my fieldwork, I did not find substantial physical modifications or adaptations in Na Maria's kitchen. However, for Maria the most significant changes concerned her preoccupation with who was going to inherit her knowledge—would it be her daughter-in-law? For her, household and kitchen modifications were not as important as the techniques and knowledge required to continue the tradition.

Regional cuisine: *Fiesta* dishes and techniques in communitarian kitchens

At the beginning of this chapter I suggested that regional cuisine requires specialized cooking procedures that emerge among some social groups in a specific physical and social region. However, affective and discursive narratives about what people believe and consider part of the cuisine on the Isthmus of Tehuantepec take into account practical and material dimensions (techniques and technologies). Na Maria's experience is only one response to these conditions that reveals everyday uses of techniques and technologies and the relationships forged within the household kitchen.

There is another arena in which we can find these procedures and instruments, but in community kitchens the relationship between cook and space is more complex. Lifecycle and community rituals are important in Zapotec communities because in these social spaces food works as a symbolic vehicle that establishes committed relationships and helps to continue traditions. I address two aspects: first, the complexity of community kitchens and the required knowledge of culinary practices; and second, the symbolic power of food prepared for ritual ceremonies. In this sense, because of their diversity, richness, and abundance—as well as because of the diversity of ingredients, time, and skills needed for their preparation—members of the community consider dishes served at *fiestas* to be the most representative of *Istmeño* cuisine.

Na Cecilia is another renowned traditional cook in the area. She was the daughter of the first traditional cook to be paid for preparing food for lifecycle and community ceremonies. In other words, Cecilia inherited knowledge, techniques, technologies, and social relationships from her mother. In Zapotec communities, social relationships[10] are fundamental for achieving recognition as a traditional cook. When her mother died, she proudly took over the responsibility of her profession. When the most important local *fiesta* is celebrated, the *mayordoma* (the woman who organizes the *fiesta*) hires Cecilia to make enough food for eight to ten days. From the outset, Cecilia will call in some extra help (her daughter, sisters and *comadres*) in order to finish the job on time.[11] The community kitchen is built where the cooks need it, and the *mayordoma* provides the necessary material and food supplies requested by Cecilia. She rents pans, casseroles, knives, buckets, ladles, and a refrigerator, as well as tables and chairs so that the cook can prepare food for people who attend the celebration. Cecilia is in charge of orchestrating the whole food preparation process. She instructs helpers (to wash, chop, grind, and serve food), sets the fire to the correct temperature, and seasons dishes in her community kitchen. Space size, design, and accommodation vary from time to time, but the elements remain broadly the same. There is one place to store food supplies (vegetables, fruits, and spices), another to store cooking instruments (pots and cutlery), another for the refrigerator and beer coolers (served by men), and an area for wooden tables and chairs. Poles to support the roof are also necessary because the cook and helpers hang their hammocks there to sleep in the kitchen after the working day is done.

There are three main categories of food prepared, served, and consumed during these *fiestas*. The first is the everyday food for organizers (like coffee, white *atole*,[12] sweet bread, *frito de res, tortillas*, and *moronga*). There are also snacks (*botanas*) that consist of small portions of salty peanuts, potato chips, shrimp cocktail, cooked crab, bean tamales, turtle eggs, mashed potatoes, carrot and potato salad, fried fish, and *getabingi*[13] that the cooks serve on individual plates. Finally, the menu includes regional dishes such as black *mole*, red rice and *barbacoa*. The everyday food and regional dishes are the responsibility of the cook, while the snacks depend on the relationships that the *mayordoma* has in the community. Sweet beverages and cold beers accompany everything.

Food preparation requires many people's help and it is therefore important that the *mayordoma* and the cook combine efforts and networks so that the *fiesta* will be a success. Sometimes Cecilia has to organize more than 100 women at once to prepare *tamales de mareña* (all of the ingredients are raw and wrapped in banana leaves). The community kitchen is transformed into a large learning space. Some women cut banana leaves into squares with their hands, others use spoons to make maize dough balls, others prepare the meat and sauce with their hands and spoons, and others fold and tie up the *tamales* and put them into a plastic container before taking them over to the cook who finally puts them in the pot to steam. Very clearly women have a variety of roles in community kitchens and cooks must have a great deal of experience to organize large-scale catering events.

I am unfortunately unable to address the historical dimension of regional cuisine in this chapter. Regional cuisine is constituted by economic and symbolic goods and articulated in collective knowledge and beliefs about the representative regional dishes that involve specialized procedures for cooking. Social and material spaces limit this collective knowledge and beliefs to the women connected with the kitchen of the Isthmus. The ongoing use of the "trial and error" technique by traditional cooks and apprentices in different ritual scenarios in community kitchens allows group members to rethink their ethnicity. Therefore, Zapotec cooks and *commensals* understand regional cuisine as a cultural patrimony, a shared legacy embodied in the personal-collective experience.

Concluding remarks: Technologies, regional cooks, and ethnic identity

This chapter demonstrates how modern and traditional technologies coexist in *Istmeño* kitchens, and also provides some ideas regarding the different practices required to produce regional cuisine in two different social spaces: the household and community kitchens. In these spaces, cooks and helpers share and continuously reinforce ideas about the foods that represent regional cuisine in different rituals. However, as demonstrated throughout the chapter, women are constantly examining, negotiating, and upgrading these ideas in social spaces. I have emphasized the relationships that generations and gender differences establish in these spaces.

Nowadays, home cooks and family members are transforming traditional rural kitchens as migration fluxes occur. However, transformations of space and ideas are not new, as it becomes evident when we use a historical approach for their analysis. For example, Zapotec women have always been able to adopt and adapt new technologies and create new techniques in which tradition and modernity join forces to engender new dishes (as evinced by the diversity of *botanas*). As suggested by Maria's story, some kitchen adaptations are possible following negotiations between family members. This is more difficult in community kitchens because many people are involved, but they remain the main cook's domain. Rural household and communitarian kitchens can be located in different spaces, such as streets, houses, private rented places, flea markets, and even on the beach. In this chapter, I described two of them to illustrate how women use and modify technologies and techniques depending on the space and time available to them.

The ethnographic method challenges hegemonic modern culinary values (perfection, sophistication, purity, hygiene, and design), which describe technologies, techniques, and physical spaces for food production. In this chapter, I suggested that negotiations are required among the Zapotec people in Oaxaca as they seek a balance between modern and traditional values. In this case, Zapotec women seek to affirm their ethnic identity through their cooking practices, and at the same time introduce changes while appropriating new technologies or ingredients into their traditional everyday cooking. Household, commerce, and *fiestas* are gendered arenas where women are visible and have a predominant role. Therefore, how technology and techniques are used and modified can be understood through a gendered lens. Men and women also develop, transmit, and practice representations of cuisine, but this chapter does not discuss that particular area.

To conclude, regional cuisine is constituted by economic products and relationships as well as the symbolic goods articulated in collective knowledge and beliefs. Cultural practices allow members of an ethnic group to re-think and negotiate their ethnicity based on a long historical process that establishes what type of food or dishes belong to the cuisine. In this chapter I also showed that Zapotec cuisine is located in local and regional spheres, but do not exclude the possibility that people think and practice regional cuisine in other spaces. In fact, the power of food transcends space dimensions because it generates affectivity: it evokes feelings (Ayora-Diaz 2014; Pérez 2014).

Notes

1 All names in this chapter are pseudonyms. Informants' names were changed for confidentiality purposes. The ethnographic material presented in this paper was collected during fieldwork between 2006 and 2009, at a small town located on the eastern part of the Isthmus of Tehuantepec, in the state of Oaxaca. My research was funded by a grant from the National Council for Sciences and Technology (CONACYT for its initials in Spanish), awarded during my graduate studies in Social Anthropology at the *Colegio de Michoacán, A.C.*

2 As a concept, *mestizaje* forms part of the early twentieth century nationalist ideology that explained Mexican culture and society as arising from the mixture of indigenous people and Spanish conquerors.

3 It also appears in cookbooks written by native and foreign authors (Guzmán 1982; Trilling 2003; and Kennedy 2009).

4 "Ethnicity" is a term that has been long discussed (Hutchinson and Smith 1996). As I argue elsewhere (Magaña González 2012), ethnicity refers to the sense of belonging to an imagined and/or real community that results from a long historical process in which interethnic power relationships can be found at local and regional levels. Hence, ethnic groups produce symbolic spaces (such as kitchens) where it is possible to analyze the negotiations and conflicts that emerge and allow people to feel part of a larger community.

5 The *totopo* is a hard *tortilla* 7 to 15 cm in diameter. It is usually made with *zapalote chico* maize, a variety developed by Zapotecs nearly 1,000 years ago. *Gendarmes* are a type of hard maize *tortilla* with an oval shape were meat or vegetable toppings can be added.

6 *Guaje* is the local term that refers to a genus *Leucaena* that grows on trees. It is often used to store and keep the *tortillas* warm. The *pichancha* is a *guaje* perforated to use as sieve.

7 I develop the concept of "societies of prestige" elsewhere (Magaña González 2007, 2012). In those texts, taking a *longue durée* approach, I explain how symbols of prestige change through time. However, the value and performance of ethnicity, or belonging to a family or ethnic community, is a value that persists in Zapotec discourse and is continuously re-elaborated through political, economical and cultural practices.

8 "Na-" is a prefix commonly used to respectfully address elderly women. The prefix "Ta-" is used to address men with the same connotation.

9 *Ayudas* ("helps" in English) is a traditional reciprocal practice that refers to *tequio* or in classic anthropological literature such as Malinowski's *Kula* (1967) or Marcel Mauss' *The Gift* (1990). *Ayudas* are economical arrangements and social consensus that establishes that when a person received economical or physical assistance, in another moment he or she has to repay the "help." received. Along, the economic relevance, the places where this economic system is developed, sustained and negotiated during life cycle and communitarian (religious and political) rituals.

10 Kin relationships and godfather–parent relationships (*compadrazgo* and *comadrazgo*) are essential for almost every activity in the community, especially those related to politics and economic decisions. These relationships allow community members to maintain a sense of belonging to develop relationships, and are part of local identity.

11 For Cecilia, her job—in addition to its value as economic reciprocity—involves a great commitment to the people who invited (hired) her.

12 A traditional hot corn and *masa* (dough) based beverage made with grounded maize, water, sugar, and other ingredients.

13 A shrimp and maize croquette served on special occasions.

References

Ayora-Diaz, S. I. (2014), "Estética y tecnología en la cocina meridana." Pp. 317–341 in S. I. Ayora-Diaz, and G. Vargas-Cetina (eds), *Estética y poder en la ciencia y tecnología. Acercamientos multidisciplinarios*. Mérida: Ediciones de la Universidad Autónoma de Yucatán.

Coronado Malagón, M. (2009), "Las viajeras zapotecas del Istmo de Tehuantepec." *Cuadernos del Sur* 14(27):53–67.

Coronado Malagón, M. (2000), "Los apodos de la resistencia: estereotipos, gentiliciosa zapotecas en el Istmo de Tehuantepec." *Alteridades* 10(19):79–88.

Goody, J. (1995), *Cocina, cuisine y clase: Estudio de sociología comparada.* Barcelona: Gedisa.

Guzmán de Vasquez Colmenares, A. M. (1982), *Tradiciones gastronómicas oaxaqueñas.* Oaxaca: Comité Organizador del CDL Aniversario, Oax.

Henestrosa, A. (2000), *Recetario Zapoteco del Istmo: Cocina Indígena y Popular.* Mexico City: CONACULTA-Culturas Populares, no. 33.

Hutchinson, J. and Smith, A. D. (eds) (1996), *Ethnicity.* New York: Oxford University Press.

Kennedy, D. (2009), *Oaxaca al gusto.* Mexico City: Plaza y Janes.

Lévi-Strauss, C. (1964), *The Raw and the Cooked.* London: Cape, Weightman J & D.

Lévi-Strauss, C. (1965), "Le triangle culinaire." *L'Arc* 26:19–29.

Magaña González, C. R. (2007), "Somos simplemente Ixhautecos: Una mirada a la identidad local de San Francisco Ixhuatána, Oaxaca." Master's thesis in Social Anthropology. Zamora: El Colegio de Michoacán.

Magaña González, C. R. (2012), "Etnicidades en construcción: Procesos históricos y relaciones interétnicas entre huaves y zapotecos en el oriente del istmo oaxaqueño durante el siglo XX." PhD thesis in Social Anthropology. Zamora: El Colegio de Michoacán.

Malinowski, B. (1967), *Argonauts of the Western Pacific.* New York: E.P. Dutton.

Mauss, M. (1990), *The Gift: The Form and Reason for Exchange in Archaic Societies.* New York: Norton and Company.

Millán, S. (2005), "Identidades vecinas. Relaciones Interétnicas en el Istmo de Tehuantepec." Pp. 141–170 in M. Bartolomé (ed.), *Visiones de la diversidad. Relaciones interétnicas e identidades indígenas en el México actual.* Vol. II. Mexico City: ENAH.

Pérez, R. L. (2014), "TecnoEstética translocal y alimento en MexAmérica." Pp. 299–316 in S. I. Ayora-Diaz and G. Vargas-Cetina (eds), *Estética y poder en la ciencia y la tecnología: Acercamientos multidisciplinarios.* Mérida: Ediciones de la Universidad Autónoma de Yucatán.

Reina Aoyama, L. (1997), "Las zapotecas del Istmo de Tehuantepec en la reelaboración de la identidad étnica, Siglo XIX." Paper presented at the XX LASA International Conference, 17–19 de April, Guadalajara, México.

Scott, J. C. (1985), *Weapons of the Weak: Everyday Forms of Peasant Resistance.* New Haven, CT: Yale University Press.

Trilling, S. (2003), *Oaxaca sazón de mi corazón. Un gran recorrido culinario.* Mexico City: Diana.

Tutino, J. (1980), "Rebelión indígena en Tehuantepec." *Cuadernos Políticos* 24:89–101.

PART TWO

Transnational and translocal meanings

5

The Americanization of Mexican food and change in cooking techniques, technologies, and knowledge

Margarita Calleja Pinedo, Universidad de Guadalajara

Introduction

Based on my examination of the documents comprised in the collection *Feeding America* (https://www.lib.msu.edu/feedingamericadata/), I analyze in this chapter the culinary uses of chili peppers in the United States and how the Mexican dish called *carne con chile* became Americanized. The path towards Americanization encompassed three distinct processes:

1   the introduction of chili peppers into American cookery;

2   the transit of *chile con carne* from a homemade dish into an industrialized food product destined for the market that took place in Texas during the late nineteenth century;

3   the introduction of technological innovations and appliances into home kitchens, which changed cooking procedures and ingredients and led to the transformation of this dish.

The analysis of these changes reflects on how food culture, food technology, and society are intertwined.

The term "cuisine" refers to the transformation of ingredients into entirely different creations, while the use of artifacts to manipulate and preserve food by means of fire, cooking, drying, smoking, or salting implies a fundamental change in the natural condition of the ingredients used. Food, therefore, is culture in the act of cooking (Montanari 2006: 29). Here I understand cooking as a practical art that is collectively exercised; one in which technology refers to an organized knowledge of cooking that encompasses the practical rules and methods used to achieve a given end (Schatzberg 2006). But cooking also involves mental and cognitive processes that require perception and memory of the world surrounding human beings so that they are able to select the elements of nature to which actions of cooking and eating are directed. In addition, cooking is a communicative process through which experiences and knowledge are transmitted within the group and passed through generations. In this sense, cooking has a shared memory that evokes experiences and meanings that refer to the past (Holtzman 2006).

From the perspective of the sociology of knowledge (Berger and Luckmann 1966), it can be argued that cooking and its products—dishes, meals—constitute a socially constructed reality that is perceived, created, learned, and legitimized within a given society. The act of cooking food is the result of repeated actions in everyday life that become institutionalized within the social group and eventually become part of the logical, normal, practical rules and methods for preparing dishes. Individuals in the social group socialize, internalize, and externalize the culinary knowledge and culinary repertoire which appears to them as a human product that exists prior to, and independently of, their existence as a factual, objective reality. As a result, individuals appropriate the flavors, skills, and knowledge that are then played out in their acts of cooking and eating. In other words, individuals construct their cooking and eating reality through the day-to-day social interaction between individuals and society, a phenomenon limited to the specific society that creates them, endows them with subjective meaning, and structures their culinary identity in a specific sociohistorical context.

From the sociology of knowledge approach (Berger and Luckmann 1966), I consider cookbooks to be reservoirs of accumulated knowledge, meanings, and experiences that can be preserved over time and transmitted to future generations. The names of dishes, ingredients, procedures, and utensils constitute a form of typified knowledge that represents an objectified reality, a codified knowledge that individuals appropriate and externalize in the act of elaborating and/or following recipes. Although published recipes seem unchanged over time, when compared they reflect changes in the technological, physical, and material conditions of food, the availability of ingredients, and cooking procedures as food production becomes industrialized and inserted into commodity chains. Finally, such modifications bring about changes in cooking times and the socio-cognitive framework of how a group and individuals think, prepare, and consume food.

My goal in this chapter, then, is to examine the historical, social, and food-mechanization processes that changed Mexico's *carne con chile* into what is known in the U.S. as *chile con carne*. On the one hand, I show how the organized knowledge of cooking codified in, and transmitted through, written recipes has changed over time; on the other, I describe how a culinary patrimony was preserved through the use of ancient technology, while analyzing the oral communication of recipes among early Mexican residents in the United States.

Early use of chili

Archeological evidence shows that chili (*Capsicum annuum*) was domesticated in central Mexico between 7000 and 5000 BC. From Mexico, Columbus introduced chili to Spain, and Portuguese traders soon extended it to Africa and Asia in the sixteenth century (Crosby 1991). The variety of chili introduced into India was of Mexican origin, although much later the culinary use of chili in Indian spice mixture (curry) traveled back to the American continent through New England's colonies. The Spaniards, meanwhile, introduced the use of raw Mexican chili peppers into the southwestern territory of what is now the United States, but which at that time still belonged to New Spain (Long-Solis 1986). A review of an online collection of seventy-six cookbooks published in the United States between 1798 and 1922 from the *Feeding America Project* at the University of Michigan provides testimonies of both these influences.

The British influence regarding chili in cookbooks published in the United States can be perceived in the inclusion of curry powder in twenty-four recipe books from the collection, which together include over 100 curry-based recipes. Few cookbooks indicate the procedure followed to prepare curry powder from scratch (which requires using a mortar to blend the ingredients), so most of those recipes list it as a preprepared spice called "Indian curry powder," which could be found readily in first-class grocery stores and pharmacies (Gillette 1887: 81). It was on these terms that curry powder was added into the dishes, although I found one recipe that explains how to bottle cayenne pepper, one ingredient in curry.

> Take ripe chilies and dry them for an entire day before the fire, turning them frequently. When quite dry, trim off the stalks and grind the pods in a mortar until they become a fine powder, mixing in about one-sixth of their weight in salt. Or you may grind them in a very fine mill. While grinding the

chilies, wear glasses so that your eyes will not be irritated by them. Put the powder into small bottles, and secure the corks.

LESLIE 1840: 182

Vinegar represented another use of capsicum, as it was made with "a hundred red *chilies* or capsicums, freshly-gathered, cut into small pieces and infused for a fortnight in a quart of the best vinegar, shaking the bottle every day, then strained" (Leslie 1840: 180).

In the United States, chili pods were available in the markets of northeastern cities where two varieties of pepper or capsicum could be found: the hot pods that are initially green in color but turn red when ripe; and bell peppers, which are the least pungent. Both varieties are cultivated in the northeast of the United States and abound by September (de Voe 1867: 341). Chili peppers were also grown in domestic gardens, and imported into the United States, as demonstrated by a list of import duties dated in 1842.

In sum, in the northeastern region of the United States, chili entered kitchens as a spice or condiment rather than as a vegetable that could be cooked into a dish. Cookbook authors from that time were chefs, columnists in women magazines, and teachers in culinary arts schools who transmitted their knowledge through their publications, some of which achieved several editions over the years. Many were popular among the English-speaking population of the United States, which in 1840 numbered over 17 million inhabitants across twenty-seven states (U.S. Census Bureau). However, the use of chili vinegar was not widely embraced, as illustrated by the fact that it is not mentioned in more recent cookbooks and commercial interest was too limited to expand its use. Curry, in contrast, passed the test of time. In the *Feeding American Project's* digital database, there are six cookbooks published in the early twentieth century that offer sixteen curry recipes. In stark contrast to the diffusion of chili through printed cookbooks addressed to an anonymous public, the culinary influence of chili in Texas was marked by ongoing cultural contact between Mexican settlers and white Anglophone pioneers.

Food culture and technology in early Texas

When the Spanish colonists expanded into Texas in the eighteenth century, they found an arid land inhabited mostly by seminomadic bands of belligerent Indians devoted to hunting and gathering. In order to expand and protect its northern boundary, the government of New Spain established garrisons and missions in Texas. The first expedition of solders, priests, and hispanicized Tlaxtaltecan Indians departed from Coahuila on a mission to teach the "savage" Apache Indians to live a sedentary life. That expedition included supplies of seeds, corn, cattle, sheep, and goats, plus construction tools and cooking utensils. A blending of foods emerged from this coexistence of Indians with Spaniards, and some Mexican dishes eventually became popular in Texas, including *chocolate, atole, elotes* (corn on the cob), *chile, tamales,* and *tortillas*. The use of Indian pottery, baskets, and tablecloths was common in homes there, alongside articles imported from China, England, and Mexico. Indeed, even prosperous Spanish homes owned *metates* and *comales,* and some families even learned to scoop up their food with *tortillas* (Weber 1992: 317).

In 1820, just one year before Mexico gained its independence from Spain, the vast territory of Texas had a non-native population of only 2,000 inhabitants distributed over three settlements. San Antonio was the largest community, one where Spanish descendants accounted for 55 percent of the population, while 13 percent were Indian, and 32 percent of mixed race (de la Teja and Wheat 1985: 11). However, between 1825 and 1830, some 9,000 new Anglo and European immigrant families arrived there, soon outnumbering the Hispanic group—*Tejanos*—that became a linguistic and cultural minority. Although the presence of these newcomers invigorated the state's economy, Texas remained a basically rural society.

In the United States, fireplaces were located inside the houses, with iron pots, kettles, frying pans, trivets, and utensils such as long-handled forks, skewers, skimmers, and ladles all hung nearby. The fireplace, or hearth, was used for cooking and to provide warmth long before the cast-iron stove, that burned wood or coal, arrived in mid-nineteenth century (Plante 1995). A table placed in the kitchen was used for eating, sewing, or working. The staple food was corn, which was ground, boiled, or fried with the use of rudimentary cooking utensils such as mortars, pots, and pans. In Texas, livestock thrived on the extensive grasslands, but agriculture allowed the inhabitants to be self-sufficient. Wheat was not grown locally, so maize, or Indian corn, provided the main caloric intake for both the Anglo and Hispanic populations. Despite this fact, during the early period in Texas, the different food cultures of the two groups co-existed, remaining separate due to social segregation.

In 1849 a German immigrant in Texas wrote about foodways there:

> Rich and poor alike eat the same kind of food: Morning—meat, cornbread and coffee (maybe some potatoes and eggs); Noon—the same (except well-off people might have rolls from fine flour with apple cuttings); Evening—the same (except milk instead of coffee—some deviate by leaving out meat—sometimes they leave it out in the morning, too). In

winter there is pork. In summer there are all kinds of vegetables and rarely any soup. Even though the food is plain there is a lot to do in the kitchen because cornbread is baked in a saucepan for every meal. It is quite a nuisance, but cornbread dries out too quickly and doesn't taste if not fresh ... Everyone grinds their own corn and sifts it and the powder is what is used to make the bread. Butter is churned daily.

CAL GROSS, in Lindemann 1986: 10

In contrast, rural women in Mexico ground the corn at home, kneeling over their *metates* and setting a bonfire on the ground inside or close to their *jacal* (house made of straw and mud). A *comal* (griddle—in this case metal) was placed over the bonfire to cook the *tortillas*. Numerous documents describe the *Tejano* diet during the second half of the 1800s, which was based on *tortillas*, beans (*frijoles*), and chili, and so did not differ greatly from the typical diet of central Mexico. Foreign travelers in both Texas and Mexico mentioned the use of corn, beans, and chili in Mexican foodways, as well as food habits found further north, all of which were different, unknown, strange and unfamiliar to them.

In 1828, H. G. Ward wrote about chili or capsicum, which:

[C]onstitutes one of the necessaries [*sic*] of life with the Indian and Mestizo population and is used in very large quantities at the table of the Creoles of all ranks ... It is used by the lower classes as a seasoning to the insipid tortilla, and the two together furnish a meal, which they would not exchange for an allowance of meat, and wheaten bread.

WARD 1828: 54

In 1852, Clarke wrote about his journey through Chihuahua:

The articles of food most common in Mexico, and which soon become necessaries [*sic*] of life to a foreigner, are frejoles [*sic*], chile, tortillas, penole [*sic*] and atole. The frejole is a large brown bean, which is very rich, and everywhere used. It is generally boiled, and afterwards stewed in lard, and sometimes with a good sprinkling of chile.

The chile (red pepper) is liberally mixed with nearly all their dishes, meat and vegetable, and no true Mexican will pass the day without a dish of the genuine article itself. Even foreigners, who at first cannot endure its smart, soon become fond of its stimulant ... Tortillas are everywhere used, frequently with a rich gravy of chili, and with atole, a gruel made of penole. This they eat without knife or spoon; rolling the tortillas into a scoop, they dip up the liquid with admirable skill, and then devour the spoon itself.

(p. 52)

In summary, in Texas, Anglos and Mexicans shared the same territory and cultivated corn for their everyday survival, but held on to distinct cooking and eating cultures. From Native North American Indians, the Anglos adopted corn and learned how to cook corn kernels into edible foods. In the mid-nineteenth century, recipe books in the United States were written by individuals eager to communicate, preserve, and transmit their culinary knowledge. In contrast, because illiteracy rates were high in both the *Tejano* and Mexican rural societies, young girls learned cooking skills under the watchful eyes of their mothers and by constantly practicing the tasks of grinding corn on the *metate* and making *tortillas* on the *comal*. Evidence indicates that very few Anglo women in Texas had *metates* (de la Teja 1998), while inventories of the largely metal cooking utensils available in the US in the late nineteenth century do not include any foil sheets or griddles that bear any resemblance to the *comal* (Campbell 1978). This suggests that while culinary exchange between Anglos and Mexicans did not occur in the case of methods of corn preparation, there is ample evidence showing that chili was widely used in Texas from very early times. For example, the *Emigrant's Guide to Promote the Texas Republic*, written in 1840, mentions that chili grew wild in the territory of Texas, but that cultivated varieties grown in different colors and sizes and with varying degrees of spiciness had become a popular seasoning for meats and other dishes. This explains how it became a trade commodity, for its consumption had increased among both Mexican and Anglo settlers (Emigrant Texas 1840: 160).

San Antonio's "Chili Queens": From homemade food to street vendors

San Antonio's location on the southernmost west–east rail line, and on a north–south railway—as well as its position in the south-central area of Texas, where two-thirds of all Texas fresh fruit and vegetables were grown—made it an ideal distribution center for agricultural products. During winter, produce-buyer representatives from the northern states would meet in San Antonio to purchase from growers in the Rio Grande area and then pack and ship their goods northwards. In San Antonio, food marketing took place in the Military Plaza, where stalls were set up every morning. At night, that plaza filled with food stands attended by Mexican women who arranged tables, benches, and lanterns and lit fires to keep warm the homemade foods they sold (Noonan Guerra 1988: 10–28). Homemade *tamales*, *chili con carne*, refried beans in green chili sauce, *enchiladas*, coffee, and hot chocolate were offered every day, but *chili con carne* soon became the most popular, and famous, dish.

Carne con chile, is a generic Mexican dish made with diced pork or beef cooked in chili sauce—*salsa*—another generic dish in Mexico that requires two main ingredients: tomatoes and raw, dried, or fresh chili peppers. But from this simple recipe the multiple procedures followed in cooking *salsas* and roasting or boiling the chili peppers, plus the almost endless varieties of chili used, and their many combinations with either green or red tomatoes, provide a plethora of different flavors and aromas. Knowledge and experience with *salsas* are family and community oral traditions learned along with the use of domestic technologies that employ the *metate* and *molcajete* (a small round mortar also made of stone) to grind up the ingredients and make the salsas. In Texas, Mexican women experienced with the varieties of chili pods they selected and added their personal touch—*sazón* in Spanish—to distinguish the pots of *chili con carne* they offered for sale in the Plaza. This is probably the reason why the women who sold *chili con carne* in those food stands came to be called the "Chili Queens." There could be no one, single queen, for each one had her own special *sazón*. Although gathered in the same square, they offered a tempting assortment of different flavors that exalted the appreciation of chili.

In fact, San Antonio's chili stands became so popular that at the 1893 Chicago World Columbus Exposition, Texas exhibited a fully fledged chili stand for the whole world to see, one complete with its Chili Queens. But because Mexico was not present at that grand exhibition, which received nearly 26 million visitors, the chili stands became a reference point to those who experienced their first encounter with Mexican food, and Texas became a reference for Anglos' social construction of Mexican food. In order to raise funds to pay for the trip of women visitors who could not otherwise afford to visit the exhibition, the Board of Women[1] that participated in organizing the World's Fair in Chicago printed a cookbook in 1893. That cookbook, entitled *Favorite Dishes*, contained over 300 autographed recipes, including one for *tamales de chile* and another for *tamales de dulce* by Mrs. Manuel Chaves from New Mexico. Both of those recipes are written in Spanish and their procedures clearly assume that readers already know how to transform corn into *nixtamal* using a *metate*, and how to make the chili sauce needed to prepare the tamales, because they offer no instructions for those processes. Instead, they explain only how to fill corn leaves with the mixture (dough) and then boil them (Shuman 1893). On the other hand, the recipe for *Mexican enchiladas* written in English by Mrs. Frank Albright from New Mexico explains that:

1. To make tortillas for the enchiladas, take one quart of blue corn meal mixed with water and salt, making a batter stiff enough to flatten out into round cakes, and bake on bare hot lid. 2. To make the chili sauce: One cup

of tepid water; three tablespoonfuls of ground chili; let boil down to a batter. 3. Filling for tortillas: Grated cheese and chopped onions, very fine. Dip into a pan of boiling hot lard one tortilla; then dip this tortilla into the chili batter; then sprinkle with the filling, first the cheese and the onion. Then put in one spoonful of chili batter and lay like a layer cake as many cakes as desired, and then pour over the chili batter. Cut like cake and serve hot.

(p. 110)

This procedure for preparing enchiladas shows an Anglo interpretation of the Mexican dish in which cornmeal is substituted for *nixtamalized* corn, and chili powder is used instead of chili peppers. Indeed, of the seventeen recipes for chili sauce in the *Feeding America* cookbook collection, only Corson (1886: 323) explains how to make chili sauce from scratch. Corson's recipe calls for five pounds of fresh red chili pods (weighed after the seeds are removed, and finely chopped), five pounds of ripe tomatoes (peeled and weighed), one pound of onions (peeled and sliced), one fourth of a cup of brown sugar, two tablespoons of salt, and a pint of vinegar. The ingredients are poured into a porcelain-lined kettle, boiled slowly and stirred frequently until they thicken to the consistency of applesauce. After that, the sauce is allowed to cool in the kettle before being poured into bottles and corked tightly.

The rest of those recipes simply call for a tablespoon of sauce. Some indicate the use of chili powder or suggest that cayenne peppers should be chopped up and added into the dish when cooking soups, meats, oyster, crab or shrimp cocktails, or for salad dressing. There are four recipes for *chili con carne* in the *Feeding America* collection that date from 1905 to 1919, but they all call for chili powder, which by that time was a readily available industrialized product. Those recipes give no further instructions for the use of chili or how to prepare chili powder. At the turn of the twentieth century, the canning industry standardized the *sazón* of the Chili Queens, but when San Antonio's 1943 Sanitary Policy prohibited the street food trade the chili stands quickly went out of business (Pilcher 2008).

From street food vendors to food canning

During the economic effervescence of Texas in the 1890s, a German immigrant, Willie Gebhardt—who years earlier had established in the town of New Braunfels a modest restaurant that served *chili con carne* to his compatriots—became active in importing and commercializing peppers from Mexico. After a time, however, he became exasperated by the seasonal growing cycle and the shortness of supplies, so he decided to industrialize the variety known as *chile ancho* ("wide" chili). Gebhardt began work in his home using the trial and error method. Instead of a *metate,* he used a domestic meat grinder to crush the chili peppers several times and then mix them with other spices commonly used in Mexico, such as cumin, oregano, and garlic, among others. In 1896, Gebhardt patented his *Eagle Brand Chili Powder* and created a company called Gebhardt Mexican Foods, which soon began to receive railroad cars filled with *chile ancho* from Mexico. In 1908, his company published a promotional recipe book in English entitled *Mexican Cooking: The First Mexican-American Cookbook.* It was dedicated to American housewives "who do not know the right mixes of ingredients." As one would expect, all the recipes call for Gebhardt's chili powder. Also, in the *Feeding America* cookbook collection, *The Neighborhood Cook Book,* published in Portland in 1914 by the Council of Jewish Women, the recipe for "Short Ribs, or Spanish Beef" calls for a can of tomatoes and Gebhardt's chili powder as a substitute for paprika or red chili (p. 119), a clear indication that new knowledge and practices were being incorporated into American kitchens. In 1911, Gebhardt began a canning operation to market *chili con carne* as a ready-to-serve main meat dish. To achieve this, his firm hired Mexican women for the production lines and used Mexican recipes, the handwritten originals of which—in Spanish—are in the company's archives at the University of Texas at San Antonio.

Founded by T. B. Walker, a second chili powder company opened in Austin to produce and market *Mexene chile* powder as seasoning. In 1905 it became the first factory to can *chile con carne.* In 1914 the firm was restructured and Walker's Austex Chili Company emerged to offer several Mexican specialties: *Mexene chile* powder, *chili con carne, tamales,* and beans in chili gravy. By the 1950s both of these chili-powder producers had expanded their workforces, added new products, increased sales, and integrated automation to modernize their productive processes. In a promotional *Recipe Booklet,* Austex described itself as "a remarkable example of modern scientific methods applied to the manufacture of food products," which entailed the use of meat grinders capable of processing over 10,000 pounds of meat daily and six 250-gallon capacity aluminum kettles in which the meat and beans were cooked. However, it still required a crew of fifty women to hand-pick the peppers, remove the stems, and separate foreign material from the raw Mexican peppers before sending them on to the machinery that cut the pods open (Austex, *Recipe Booklet* [n/d]).

Sahrah McClendon, a Washington newspaper correspondent, wrote an article under the headline, "Austex *Chili* makes up main dish at meal in Washington for Texans," in which she explained that a dinner, for some 450 people held at the National Press Club including all the members of the Texas delegation in Congress and other top Congressional leaders, was a promotional strategy planned by Austex as part of its efforts to regain the eastern market that had declined during World War II when the firm was obliged to sell its products to the armed forces. At the dinner, Austex's general sales manager and dietitian explained:

> Lots of folks don't know how to prepare chili to get the real flavor out of it [canned chili] . . . Part of our job of being here tonight is to teach the people in the country how to prepare chili best—the housewife sometimes thinks it is a quick and easy dish, so what does she do? She lets it get too hot and burn while she's setting the table. Then she stirs the burn into the chili and lets the grease and water accumulate on top. Chili must be heated just ever so slowly and delicately to get the flavor out of the oil in the beans. And our chili is not too greasy and does not have too much garlic in it. We think it could become a much more popular dish throughout the nation if folks knew just how to prepare it properly.
>
> SAHRAH McCLENDON 1955

The Americanization of Mexican food

At the turn of the twentieth century, the canning industry modernized and standardized the Mexican dish *carne con chile* in an industrialized product they called "chili con carne." Those canned products and new domestic appliances, like the electric can opener, replaced the *sazón* of the Chili Queens. However, it is important to emphasize that this process of Americanization of Mexico's *carne con chile* began with the transfer of the Mexican Chili Queens' knowledge of the ingredients, their combination thereof, and the correct proportions and procedures. *Carne con chile* was originally a generic dish that formed part of the culinary heritage of Mexican women living in San Antonio, but was capitalized on by businessmen who patented their knowledge and standardized its flavor. A second significant aspect of this process was that the manufacturing process organized in departments—selection, stemming, washing, packing, etc.—fragmented women's knowledge into separate phases, and then converted each phase into a routine task requiring repetitive movements that completely suppressed the creative aspect of cooking. Third, by fragmenting the cooking procedure the factory no longer required the criteria and expertise of cooks, but rather a labor force trained to feed the machinery and complete, or complement, its work. Moreover, it was no longer necessary to have cooks supervising the operations or even tasting the final results. Fourth, the industrialization process changed the relationship between cooks and consumers. While food cooked in the home was once shared and enjoyed *in situ*, industrialized food products depersonalized the cook–consumer relationship. Fifth and finally, consumers' memory and identification of foods were no longer associated with home cooking but with commercial trademarks.

This review of the *Feeding America* cookbooks published in the United States in the twentieth century has shown how a Mexican dish was introduced, interpreted, and then adapted to Anglo kitchens and the technological resources available in Texas, which differed markedly from those that Mexican women were using in the same period following the tradition of Mexican kitchens from the nineteenth century. It can also be concluded that before Tex-Mex food became popular in the U.S. with the spread of restaurants, Anglo homes were already acquainted with Mexican chili peppers and had adapted them as an ingredient in their cuisine, their tastes, their foods, and their technology in two ways: first, in the form of Indian curry that includes the Mexican spice capsicum which arrived in northeastern America via the English colonists; and second, as *chili con carne*, which entered America through ethnic contact between Anglo Texans and Mexicans.

Note

1 This was a board integrated by women from different states, deemed influential at the time. Among their achievements, they established the Women's Building, which was considered an important innovation at the fair as it represented different aspects of women's lives and their contribution to society. See: http://digital.lib.msu.edu/projects/cookbooks/html/books/book_45.cfmøp.

References

Austex (n/d), *Recipe Booklet. Austex Foods*, File: F2500 (1). Austin Historical Center. Austin Public Library.
Berger, P. and Luckmann, T. (1991 [1966]), *The Social Construction of Reality*. London: Penguin Books.
Campbell, L. (1978), *America in the Kitchen from Hearth to Cookstove*. Orlando, FL: House of Collectibles.
Clarke, A. B. (1852), *Travels in Mexico and California*. Boston: Wright & Hasty's Steam Press. In *Sabin Americana, 1500–1926*. Gale Digital Collections [electronic version]. Accessed through the University of Texas in Austin.
Crosby, A. (1991), *El intercambio transoceánico: Consecuencias biológicas y culturales a partir de 1492*. Mexico City: UNAM.
Corson, J. (1886), *Miss Corson's Practical American Cookery and Household Management*. New York: Dodd, Mead and Co.
Council of Jewish Women, The (1914), *The Neighborhood Cook Book*. Portland, OR: Press of Bushong and Co.

Emigrant, Texas (1840), or, *The Emigrant's Guide to the New Republic: Being the Result of Observation, Enquiry and Travel in That Beautiful Country.* New York, 1840, p. 160. In *Sabin Americana, 1500–1926.* Gale Digital Collections [electronic version]. Accessed through the University of Texas in Austin.

Feeding America: The Historic American Cookbook Dataset. East Lansing: Michigan State University Libraries Special Collections. https://www.lib.msu.edu/feedingamericadata/.

Gebhardt Chili Powder Co. (1908), *Mexican Cooking: The First Mexican–American Cookbook.* Bedford, MA: Applewood Books.

Gillette, Mrs. F. L. (1887), *The White House Cook Book.* Chicago: R. S. Peale & Company.

Holtzman, J. D. (2006), "Food and memory." *Annual Review of Anthropology* 35:361–378.

Leslie, E. (1840), *Directions for Cookery, in its Various Branches.* Philadelphia: E.L. Carey and Hart.

Lindemann, A., Lindemann, J., and Richter, W. (1986), *Historical Account of Texas Industry 1831–1986.* Texas: New Ulman Enterpise Print.

Long-Solis, J. (1986), *Capsicum y Cultura: La historia del chilli* Mexico City: Fondo de Cultura Ecómica.

McClendon, S. (1955), "Austex *Chili* makes up main dish at meal in Washington for Texans." Newspaper article archived at Austex *Foods*, File: F2500 (1). Austin Historical Center, Austin Public Library.

Montanari, M. (2006), *Food Is Culture.* New York: Columbia University Press.

Noonan Guerra, M.A. (1988), *The History of San Antonio's Market Square.* San Antonio, TX: The Alamo Press.

Pilcher, J. (2008), "Who chased out the 'Chili Queens'? Gender, Race and Urban Reform in San Antonio, Texas 1880–1943." *Food and Foodways* 16(3):173–200.

Plante, E. (1995), *The American Kitchen: 1700 to the Present.* New York: Fact on File.

Schatzberg Eric, (2006), "'Technik' Comes to America: Changing Meanings of 'Technology' before 1930." *Technology and Culture* 47:486–512.

Shuman, C. (ed.) (1893), *Favorite Dishes.* Chicago: R. R. Donnelley and Sons Co. Printers.

de la Teja, J. (1998), "Discovering the Tejano Community in 'Early' Texas." *Journal of the Early Republic* 18(1):73–98.

de la Teja, J. & Wheat, J. (1985), "Bexar: Profile of a Tejano community, 1820–1832." *The Southwestern Historical Quarterly* 89(1):7–34.

University of Michigan, *Feeding America Project* http://digital.lib.msu.edu/projects/cookbooks/index.html.

U.S. Census Bureau. History through the Decades, 1840 Fast Facts. https://www.census.gov/history/www/through_the_decades/fast_facts/1840_fast_facts.html.

de Voe, T. (1867), *The Market Assistant, Containing a Brief Description of Every Article of Human Food Sold in the Public Markets of the Cities of New York, Boston, Philadelphia, and Brooklyn.*

Ward, H. G. (1828), *Mexico in 1827,* Vol. I, London, Henry Colburn. In *Sabin Americana, 1500–1926.* Gale Digital Collections [electronic version]. Accessed through the University of Texas in Austin.

Weber, D. (1992), *The Spanish Frontier in North America.* New Haven, CT: Yale University Press.

University of Texas at San Antonio. *Gebhardt Mexican Foods Company Records*, Archives and Special Collections Library.

6

Home kitchens:
Techniques, technologies, and the transformation of culinary affectivity in Yucatán

Steffan Igor Ayora-Diaz, Universidad Autónoma de Yucatán

Introduction: Changing food, shifting affects

If you ask a Yucatecan what Yucatecan food is, he or she may be inclined to list a few or many ingredients or the dishes in which they are used, giving the impression that there is a fixed way to prepare and eat a range of food. But is this correct? In this chapter I describe and discuss how regional Yucatecan cooking is changing in the current context of global-local and translocal transformations, and how people's fondness for their "own" gastronomic "tradition" is shifting toward other cuisines, leading them to increasingly shallow attachments to the regional cuisine of Yucatán. Part of these changes, I propose, relates to the changing *technoscape*, which extends into and continually transforms all aspects of everyday life, including home kitchens and home cooks.

I understand "technologies" to include everything from the simplest instruments and tools to the more sophisticated appliances, electric and electronic, that people use to prepare their meals. I look at the different pragmatic articulations that cooks perform at home, sometimes choosing so-called "traditional" tools or instruments to cook a meal, sometimes relying exclusively on technologies perceived as "modern," and sometimes combining both. In this chapter, my concern is with the techniques, instruments, tools, and any other appliances cooks use to prepare a meal. This includes the diverse ingredients that, through the use of diverse technologies, have been processed and are industrially produced and marketed on a massive scale. Yucatán's regional cooking offers a privileged point of observation for those studying technological change and its effects on food and food preparation, particularly because local people strongly identify their ways of cooking and eating as an important part of their local and regional collective identity.

I have argued elsewhere (Ayora-Diaz 2012a) that Yucatecan gastronomy is a twentieth century invention of cookery writers and restaurateurs who distilled the broad regional culinary field (represented by regional cookbooks that include up to 600 recipes) into a reduced list of around forty dishes that have since become iconic. In the enduring negotiation between regional and national identities of people in Yucatán, the sense of place and peoplehood have rooted and naturalized local taste, and the unequivocal predilection for Yucatecan meals was, at the same time the affirmation of love for the place where these recipes were created and the people who created them, supporting the construction of a common Yucatecan identity (Tuan 1974, called this type of fragile attachment "topophilia"). Until 1970 the peninsula of Yucatán had remained partially isolated from the rest of Mexico, but the completion of a road connecting the region with the rest of the country, and the gradual transformation of the regional demographic composition led, in turn, to the transformations of the urban foodscape from that decade onwards (Ayora-Diaz 2014a). Since that time, in addition to regional Yucatecan and Syrian-Lebanese food, Yucatecans have faced an expanding foodscape, beginning with restaurants that marketed US versions of Italian and Chinese food. Today, the cities of the state of Yucatán are dotted with restaurants, small eateries, and street-food vendors specializing in the food of regions as diverse as Asia, the Caribbean, North and South America—as well as different parts of Mexico. In short, Yucatecans are now exposed to an array of foods that their parents certainly did not encounter fifty years ago.

I consider these transformations as non-lineal and non-causal. However, a number of tendencies coexist: local people are becoming increasingly exposed to foods alien to regional culture; the foreign foods they consume are often local interpretations of those tastes and dishes, but they are frequently presented and accepted as "authentic." Some seek to replicate foreign meals at home, and to that end bookshops stock a dizzying array of cookbooks across a wide variety of cuisines; regional, national, and transnational super- and hypermarkets offer ingredients from across the world; and department stores and businesses specializing in kitchen hardware and appliances make foreign utensils available to professional and domestic cooks seeking to reproduce foods from overseas culinary and gastronomic traditions. Thus the availability of processed and pre-packaged ingredients, new technologies, and cookbooks encourage the adoption of new cooking techniques, or the transformation of old ones.

Mike Michael (2006) argues that the term "technoscience" must encompass not only those instruments and technologies employed in the production of scientific knowledge, but also those that are the result of applying knowledge and technologies to those material goods that we use in our everyday life. Processed food that has been "improved" with color and flavor enhancers, and provided with a longer shelf life through the use of chemical preservers, is described positively by some consumers as "convenient," "hygienic," and "modern." It transforms everyday culinary practices and the taste for and *of* food, not to mention the affects that food mobilizes (Winson 2013). As Bentley (2014) argues, babies are socialized into enjoying the taste of processed foods early on, their parents having been specifically targeted by food manufacturing companies. Industrially produced food, then, becomes inscribed as technoscience.

In this chapter I describe the transformation in cooking technologies—including changes in appliances, techniques, and ingredients—that, I suggest, find reflection in changes in taste and in the affective attachments to "traditional" regional food. As will become evident, these changes are often associated with an ambiguous and ambivalent understanding of "modernity" and "tradition" and the ways these are expressed in the preparation, distribution, and consumption of everyday meals.

Technological assemblages in the global foodscape

Mérida is the capital city of the state of Yucatán, one of the three Mexican states in the Yucatán peninsula. With a population of about one million inhabitants, the percentage of immigrants from different Mexican regions and abroad is steadily growing. The Maya peoples of Yucatán witnessed and endured the Spanish conquest in the sixteenth century. Other subsequent waves of migration—from countries such as Syria, Lebanon, Italy, France, Germany, Cuba, and Colombia—have increased the diversity of customs and impacted the Spanish regional dialect. More recently, since the 1990s the region has received a new influx of immigrants from Europe, North America, and Asia (primarily Korea and China). Since the economic boom of the end of the nineteenth century fueled by the henequen plant fiber (derived from an agave native to the Yucatán peninsula), Yucatecans have opened their homes to commodities imported from the United States, the Caribbean, and Europe, and these imports gradually changed domestic culinary practices, spawning a recognizable "taste of Yucatán" (Ayora-Diaz 2012b).

The kitchen is the ideal site in which to explore the transformations of the technological assemblage[1] that supports the creation, reproduction, and dissemination of recipes, and the institution of culinary and gastronomic traditions. The Yucatecan kitchen, in particular, can be seen as a site that mediates translocal and global-local connections between cooking traditions of the world, originating first an identifiable regional gastronomy and subsequently hosting the re-creation of other national traditions within the regional foodscape. It would be misleading to assert that there is *one* "Yucatecan" kitchen. In the rural areas of Yucatán, poor peasant and indigenous people possess very simple kitchens and appliances. Unfortunately, publications about these are scant: most studies focus on the ritual and ceremonial uses of food and drink, but leave out the description of how this food, as well as everyday meals, is produced in contemporary rural homes. Everton (1991: 53–65 and 76) is an important exception, however, as he provides photographic illustration of Maya kitchens in Yucatán and their counterpart in Cancún, where many migrate in search of work. We can see that their main instruments are iron griddles, gourds, a couple of aluminum pots and casseroles, and buckets. Another important technology he illustrates is that of the ground oven or *pib*, used to cook, among other things, ceremonial meals. It is this same technology, and the associated techniques in the rural area of Yucatán that are the subject of Quintal and Quiñones Vega's essay (2013). As they discuss, the main cooking techniques among the Maya of Yucatán are boiling, frying (in pork lard), and baking in a *pib*. Most of the ingredients they use in ceremonial meals (the subject of their essay) are locally produced, even if their historical origin lies overseas (Seville oranges, onion, pork, cilantro, and chives). I propose that we need to look beyond the anthropology of rituals and folk "traditions" if we wish to understand the complex technological, economic, social, and cultural articulations shaped in the space of urban and rural domestic kitchens. Since my research has focused on the use of urban kitchens, I look at them as the site for the production and reproduction of technological assemblages.

Global technologies become relocated at the domestic level in the form of different commodities. In what follows I describe the three types of technologies that I have found in urban kitchens in Merida: cookbooks as technologies of memory and of cultural dissemination (and some times cultural colonialism); cooking appliances; and cooking ingredients, themselves the product of technological mediation. As I argue, their combined use leads to changes in cooking techniques as well. My purpose here is not to provide a comprehensive description of all kinds of technology present in Yucatecan homes, but to illustrate how cooks use different technologies in their home kitchens.

Urban kitchen spaces

The inhabitants of Mérida possess different levels of economic, social, cultural, and symbolic capital. Some were born in the city, others are rural immigrants from the state, and still others are immigrants from various Mexican regions and abroad. In the city one finds, at one extreme, micro-houses with almost no cooking space (sponsored by the state government and City Hall since the early part of this century) and at the other, large, sumptuous houses with extremely large kitchens. During 2013–2014 I conducted research in the north of the city. While mostly higher-income families live here, some lower-income neighborhoods remain in the area thanks to "social interest" (low-cost) homes that are financed either by banks or employers (García Gil, Oliva Peña and Ortiz Pech 2012). These houses have been provided with meagre cooking space: Their kitchens are often two to three meters wide by three to four meters long. They are easily filled with one fridge, one small stove, and some kitchen cabinets. There is little room to move around and the cooks often have to do their initial prep work in the dining room before cooking the meals on the stove. By contrast, the inhabitants of the wealthier neighborhoods own large kitchens, sometimes with dimensions close to five by five or even five by seven meters. Given this amount of space, they can accommodate many more kitchen appliances than their lower-paid neighbors. One young housewife, whose husband had recently acquired a "social interest" house, told me that while she was happy that the house would eventually be theirs, she was disappointed at the size of the kitchen. She showed me how little room she had to prepare the meals, and instead has to take her cutting board, knife and ingredients to the dining table to slice, chop, and marinate, and then move back into the kitchen to cook. She said, somewhat upset, that the kitchen in the house they used to rent was larger and more suited to her cooking tasks. By contrast, in a wealthier neighborhood, a woman who told me she seldom cooks showed me a clean, neat kitchen that was big enough to house a breakfast table. In addition, a couple of electric appliances stood on the counter, and there was enough space to prepare meals on the large stove in comfort.

Cookbooks

In some cases printed and electronic cookbooks have replaced—but more often only supplement—the oral transmission of technical/technological culinary knowledge and the use of handwritten notebooks filled with different recipes. In the homes of families with varying levels of income I have found printed cookbooks that describe both the technical procedures and the type of utensils required to produce elaborate Yucatecan recipes. In one home, a single mother told me that she did not own a single cookbook, just a notebook in which she had written the recipes her former mother-in-law had dictated to her. When I asked her to show me the notebook, she began taking from a drawer assorted magazines, pamphlets distributed by industrial producers of canned and other processed meals and ingredients, and small cookbooks covering pasta, meat, and grilled dishes. When I pointed out that she had all these other sources, she answered: "they are not cookbooks. Are they?" She lent me her notebook and I found it was filled with dozens of recipes copied from printed cookbooks. When I was late returning her the notebook, she told me not to worry because she had another one. Furthermore, she said she was really not in a rush to get it back because her children like only a very limited number of meals, and she already knows how to cook them.

In other homes I have found only cookbooks specialized in Yucatecan and/or Lebanese food, while in other still I have found that some cooks own collections of printed cookbooks that run from between five to forty, covering everything from Spanish tapas and pasta to Italian, French, Russian, Puerto Rican, Cuban, central Mexican, Oaxacan, Greek, and other national cuisines. When I asked one woman how often she cooked recipes from those cookbooks, she told me "never!" Her husband likes the food he eats at different restaurants and buys cookbooks for her, but she never tries to obtain the ingredients or appliances necessary to recreate these at home. She told me that she cooks Yucatecan food, and keeps the cookbooks on her bookshelves. He, on the other hand, regularly uses cookbooks to prepare special dinners for their friends. I have met other families that own fewer cookbooks, but every so often try to cook Japanese, Chinese, German, French, Italian, Cuban, or Puerto Rican meals, to name but a few. In families where the wife, the husband, or both are of Lebanese origin, the preferred recipes are usually Lebanese and Yucatecan, in either case adapted to a Yucatecan-Lebanese taste. Still, in many of these cases the daughter or daughter-in-law has learned the recipes from the mother or mother-in-law through oral instructions and trial and error. Similarly, some women who claim to cook only Yucatecan recipes have told me that they do not own any cookbooks and have learned the recipes from their own mothers, sisters, relatives, or in-laws. Some have told me that although they do not own cookbooks, they borrow from their mothers the old Yucatecan cookbooks in their possession. But this happens infrequently, as most of the meals they cook are their children's and husband's favorites. They make them week after week, and so no longer need to read the recipe.

As technologies, cookbooks inscribe in the bodily habitus and in the group's memory the procedures and the taste a meal "should" have. Cookbooks do not work in isolation, however, but rather reinforce the individual memory of meals consumed with the family or friends, and on everyday and special occasions (Sutton 2001). In addition, some cookbooks become tools aiding the collective or individual memory of migrants in (re)producing meals that they find meaningful in new social environments. That said, cookbooks may also be instrumental in imposing the hegemonic and homogenizing taste of the dominant groups of a nation upon a regional tradition, as exemplified by a woman from Central Mexico who wrote a Yucatecan cookbook adapting the recipes to their taste (Ayora-Diaz 2012a: 153–157).

Cooking appliances

These material objects are important in the transformation of ingredients into a meal that is meaningful to those who consume it. For example, Ohri (2011) describes the adaptation of kitchen spaces in the United States among Indian immigrants so that they were able to use the technology they require to reproduce the food of their country of origin. Also, Nakano (2009) has analyzed the global market for electric rice cookers among Asian migrants. It turns out that manufacturers have had to adapt their cookers to cater for the taste and texture of rice preferred in different Asian societies. Similarly, immigrants who have resettled in Yucatán have had to introduce technologies they considered essential to cook their own meals. For example, a friend told me that his Lebanese-born grandmother brought with her a large mortar in which she would patiently pound meat until it was right for making *kebbee*. Also, before the most recent wave of immigrants from China and Korea, it was near to impossible to find in Yucatán a *wok* of any kind. Now it is possible find them in aluminum, steel, cast iron, and enameled cast iron. On the other hand, some European appliances—such as stovetop Italian *moka* and French-press coffee-makers, fondue pots, clay casseroles, and paella pans—have been available for decades.

More recently, hypermarkets and department stores have made available many different appliances formerly not used by the inhabitants of Yucatán, such as French-made casseroles, grills, Dutch ovens, and other pots; manual and electric pasta presses; electric espresso machines; electric coffee grinders; electric mixers and high-power blenders; US-, German-, and Japanese-made specialty knives; stands on which Spanish and Italian meats can be sliced; Japanese mats to roll sushi; electric deep-fat fryers and grills; and electric home meat grinders (which are displacing manual grinders). The list is endless. However, in the homes where I have been interviewing home cooks, their repertoire of appliances tends to be more limited—and this is true even among the wealthy families whose diet continues to be predominantly Yucatecan.

In the kitchen of the young, recently married woman I described above, I found a small refrigerator, a small stove with four burners and oven, and some plastic shelves on which she placed four non-stick aluminum pots, three non-stick aluminum skillets, and one casserole. The couple owns a blender and a microwave oven that, she says, they do not use to cook but to reheat meals or to make popcorn. Visiting the homes of women who have been married for between fifteen and twenty years, I did not find many differences: there were perhaps a few more pots, pans, and skillets (often with non-stick surfaces), more dull knives, and perhaps an electric steamer, an electric pressure cooker, knife, or food processor, as well as ceramic or clay casseroles for Mexican or Spanish recipes. In some cases, the appliances had never been used. One woman told me that one day, her husband arrived home with an electric vegetable steamer. After several years, the appliance remains in its box—untouched. She told me: "It looks nice, but I seldom steam things, and when I cook *tamales* I need a larger container." Electric rice cookers have met a similar fate: one woman told me that after trying (and failing) to use it to make popcorn, she keeps it in the top cabinet of the kitchen, while she continues to cook rice the way she has always done.

While Yucatecan food can be elaborate and time consuming, it does not demand specialty appliances. Standard pots, skillets, casseroles, and frying pans suffice. Home cooks use bowls to marinate the meat, knives to chop, mince, or slice the ingredients, a manual grinder to ground meat for pâtés and *picadillo*, and boiling and frying remain the most common cooking techniques at most Yucatecan homes. Some recipes (such as turkey *escabeche*) require grilling the meat over coals. However, most cooks would just skip this step if they lack a grill, or roast the meat over the gas stove's burner. Most young women have distanced themselves from the kitchen and do not know how to use pressure cookers, and oftentimes are afraid to use them. Stove ovens are normally used as storage space, and given the small size of their kitchens and the high temperatures of the city (sometimes reaching, during the spring, over 45 degrees Celsius), most women or men prefer not to bake at home. In fact, some meals that need baking—such as *mucbil pollos*, a type of tamale cooked for the Day of the Dead and All Saints—are normally taken to the nearest commercial bakery. Also, only specialists cook *pib* dishes, such as chicken or pork *pibil*, as most homes would not have dug ovens in their backyards.

It is only when visiting the homes of Yucatecan couples and families that have lived abroad, immigrants who have permanently moved into Yucatán, or foodies that I have found appliances with specialized use: these include wooden mortars to prepare Caribbean *mofongo*, woks, crystal recipients for Russian caviar, pizza stones, *terrine* pots, and Dutch ovens. Many of these homes (but not all of them) have dishwashers, deep freezers, convection ovens, electric wine coolers, and sometimes even restaurant-style cooking ranges. The cooking possibilities at these homes have expanded well beyond the boundaries of regional gastronomy, and the family members tend to display shallow attachments to Yucatecan food, often favoring fried chicken, hamburgers, pizzas, and other so-called international foods over their own regional culinary traditions. Also, some families have "second kitchens" in other properties either by the sea or in the countryside, where they undertake grilling and baking. At these places it is possible to find Argentinean meat grills, Italian-style gas or wood-fired ovens, and *pibs* for them, their friends, or their domestic assistants to cook special dishes.

Ingredients

We do not normally tend to include cooking ingredients in discussions about cooking technology. However, many ingredients have become "Trojan horses" in that they are the means by which contemporary technologies enter the home and eventually become essential to our everyday lives. We may use very simple technologies to produce aromatic herbs in own gardens, or seek to purchase eggs, vegetables, cheese, or meat from organic producers who tell us they prefer not to use industrial equipment as they cultivate their crops or raise their animals. On the other hand, we are aware of the insidious ways in which genetic technologies can enter our kitchens and our daily meals depending on how we buy everyday staples that large transnational agribusinesses have transformed through the manipulation of their genetic makeup, or through the use of chemical fertilizers and pesticides (such as soy, coffee, corn, wheat, and rice). In different regions of Mexico, meat growers are often accused of using clenbuterol to fatten their animals, with toxic consequences for human consumers.

The most common vehicle for technological intervention in our everyday food remains the use of pre-cooked, processed foods (frozen, canned, boxed, bagged, or in jars). Food has been transformed in this way for many years. In fact, it was during the nineteenth century that industrialized foods began to displace older preserved foods such as salted and smoked meats (Winson 2013). During the twentieth century, the popularization of refrigerators as common household appliances (Rees 2013) added to the "naturalization" of industrialized and processed foods as part of our everyday diet. The growing presence of canned, packed, and frozen ingredients in the daily diet of the Yucatán has changed the flavor of local meals over time, also gradually altering

Yucatecans' taste for certain foods, and impacting the affective attachment to Yucatecan culinary taste and to regional gastronomy.

Until the last century, most housewives cooked at home using fresh ingredients. Markets in the city sold mainly produce, even though non-edible commodities were also on sale there. Men and women walked or drove to the market on an everyday basis, and the mother or grandmother cooked the day's meal (or in some cases, the maid/cook followed the instructions of the housewife). Thus, Yucatecans cooked regional recipes in which the main ingredients were pork, chicken, or turkey and local vegetables. The food was flavored most commonly with Seville oranges or limes, red or white onion, radish, garlic, red tomatoes, achiote, coriander seeds, cumin, cloves, cinnamon, allspice, black or white pepper, oregano, cilantro, thyme, bay leaves, saffron, epazote, chaya leaves, and *xkat* and *max* peppers. These ingredients were combined in different manners in different recipes to produce and naturalize an identifiable Yucatecan taste (Ayora-Diaz 2012a, 2012b).

In Yucatán, as in many other parts of the capitalist world, industrially processed ingredients have been available for the consumer since the late nineteenth century, but they tended to play a marginal role in everyday cooking: for example, in order to get breakfast on the table more quickly, the domestic cook would open a can or a bottle and pour the sauce on eggs or meat. For the main meal of the day, however, the home cook nearly always employed natural, fresh products. During my conversations in Mérida I found that most home cooks claimed to use very few processed ingredients—and even then only occasionally. However, when I examined (with their permission) their kitchen cabinets and refrigerators, I found frozen meats and packaged vegetables from transnational food companies, bottles of catsup, mayonnaise, and mustard, soy and English sauces, pickled olives, onions, and capers. In their cabinets they had varying numbers of canned chili peppers, refried beans, tuna and sardines, tomato sauce, *salsa casera* (canned tomato, onion, and chili pepper sauce), chipotle, and many other industrially processed and preserved foods. Yet the women I talked to insist that they seldom used them and normally cooked with fresh ingredients that they acquired at nearby supermarkets, markets, and street vendors.

However, in some other instances I found home cooks who have prepared whole meals by combining the contents of several different cans, and both they and their relatives were happy with the results (Ayora-Diaz 2014b). On these occasions, it has been the value of convenience has been the main driver in their culinary practice. For example, one young single mother told me that, since she had to work, she was happy to have found canned, pre-cooked chicken, as this made it easier for her to cook for her child. I should also say that cooking with almost exclusively processed ingredients is done privately and not normally displayed to outsiders (although on occasion I have been invited to eat some of these meals). When cooking for friends or as part of a celebration, the home cook usually tries to follow local recipes to the letter, and to use mostly fresh products—although he or she may skip one or two steps to save time.

As I have described thus far, a new cooking assemblage is emerging in Mérida, in which changes to its component parts have led to different degrees of culinary attachment. Starting in the twentieth century, local domestic kitchens have become progressively smaller; restaurants and cookbooks have exposed Yucatecans to other gastronomic traditions; different and often new cooking technologies have become available; and pre-processed, packaged, and preserved industrial food items have become ever-present in kitchen cabinets and refrigerators. In light of these changes, cooking techniques are also bound to change as part of the local culinary assemblage.

Culinary techniques

Culinary techniques are related to the other parts of the general food and cooking assemblage. Their changes find reflection in cooking techniques and, reciprocally, these techniques condition our predilection for different types of cookbooks and our preference for certain appliances, tools, and ingredients. As noted above, "traditional" Yucatecan cooking can be very elaborate. To cook turkey in *escabeche*, for example, the cook must begin by marinating the meat, then parboil, baste, and roast it over coals before adding it back to the broth along with other roasted ingredients (garlic, onion, and *xkat* chili pepper). The cook also has to make a short pasta, which will be added onto the broth, and then pickle red onion with oregano, allspice, bay leaves, and habanero pepper. Today, however, in the cramped modern kitchen or in homes with a small backyard, roasting over coals may be challenging. And let us not forget that most people work outside the home and have little free time in which to cook. A pressure cooker can be used to reduce cooking time and to skip steps, but this may mean that the taste of the meal changes in the process. Something similar happens with dishes that would ideally be cooked in a *pib*, like pork or chicken *pibil*. Nowadays they may be made either in the pressure cooker or in the oven, and while they may resemble the original versions, they are different in terms of color, flavor, aroma, and texture, and in their new guises they transform the taste of Yucatecan food for the younger generations.

If using "traditional" techniques to make *pan de cazón* (baby shark cake), the cook has to invest time in roasting tomatoes, garlic, and onion to make a sauce that will be fried along with whole habanero peppers and epazote leaves. She also has to boil black beans with onion, pork lard, and epazote, before mashing and frying them. She must in addition stew the baby shark in the tomato sauce with epazote, and then shred it by hand. If following once cooked she shreds it by hand. "Modern" techniques, however, involve opening a can of tuna, a can of tomato sauce, and a can of refried beans (or reconstituting, with water or with broth, black beans flour). In heating all the ingredients together, she makes a meal. Sometimes, analogous procedures are followed in cooking foreign foods such as pizza or pasta, since the pizza base mix (or the pizza dough itself) is available at local supermarkets along with canned "Italian" tomato sauces, flavored olive oils, and processed shredded or powdered cheese. Although some home cooks I know take pride in cooking "from scratch," most find it convenient to use at least some pre-cooked, processed and packaged ingredients that bring the taste of their meals closer to that which they have learned to appreciate at fast-food restaurants. A consequence of these changes in the taste of and for food finds reflection in the affective attachment of consumers to food in general, and to regional food in particular.

Discussion: Fluid attachments in the contemporary foodscape

Global–local and translocal intersections find expression in the movements of peoples, edible commodities, culinary techniques and technologies, and gastronomic ideologies. In the Yucatecan case, regional cooks adopted the values and taste of different foreign cuisines, and adapted them to locally available ingredients, giving birth to an identifiable and encompassing the local cuisine. The list of Yucatecan recipes was later shrunk into a short collection representative of regional taste. The naturalization of taste, along with an everyday reinforcement of ties between food, family, and social bonds, has contributed to promote a strong affective attachment to the food of the region. Its (re)production—including the techniques, tools, instruments, appliances, and ingredients—circulation, and modalities of consumption in proper etiquette strengthen the regional sense of collective identity.

However, the most recent wave of change promoted by these global–local and translocal processes corresponds to the development and dissemination of contemporary values and ideology that relativize the worth of any cultural production and encourage the recognition of difference. In this context, contemporary changes in the culinary-gastronomic assemblage and foodscape foster the emergence of fluid tastes and food preferences. The cuisines that today are taking roots in the Yucatecan foodscape are consumed by Yucatecans as well as by the foreign-born nationals who produce meals culturally meaningful to them. In some cases, Yucatecans "experiment" with imported tastes but remain faithful to their own gastronomic "tradition." However, there are other instances in which Yucatecans have come to like the flavor of industrialized ingredients and meals, driving changes into the local tradition, while others develop taste preferences for one or more imported national or regional cuisines, thereby loosening their attachment to the food of their parents. It is in the context of these processes that we need to examine how these culinary-gastronomic assemblages (that include technologies of all types) take form in urban and rural kitchens, and the effects they produce on the local and regional collective identification with Yucatecan food and gastronomy. Furthermore, I believe that understanding "technology" as encompassing tools, instruments, appliances, cookbooks, ingredients, and cooking techniques, can help us explain how its changes lead to transformations in the taste of food, and its social and cultural significance, as well as to the affective attachment to a collective identity.

Acknowledgments

This chapter is based on research conducted since 2012 with financial support from CONACYT (#156796), for a team project I direct. Some of the themes discussed here have been presented at our seminar on performance, launched by Rodrigo Díaz Cruz and Anne W. Johnson, and at the AAA meetings in Chicago, 2013. I thank all my colleagues and friends for their comments, and am grateful to Gabriela Vargas-Cetina for her critical revision of this chapter.

Note

1   I am favoring the use of "assemblages" over that of "fields" or other concepts derived from structural oppositions to convey the transitory, ephemeral, fluid, and non-stable association between parts (de Landa 2006).

References

Ayora-Diaz, S. I. (2012a), *Foodscapes, Foodfields and Identities in Yucatán*. Amsterdam: CEDLA / New York: Berghahn.
Ayora-Diaz, S. I. (2012b), "Gastronomic Inventions and the Aesthetics of Regional Food: The Naturalization of Yucatecan Taste." *Etnofoor* 24(2):57–76.
Ayora-Diaz, S. I. (2014a), "Lo Posnacional y la fragmentación del paisaje culinario yucateco: transformaciones contemporáneas en los hábitos culinarios." Pp. 49–62 in A. López-Espinoza and C. R. Magaña González (eds), *Habitos Alimentarios*. Mexico City: McGraw-Hill..

Ayora-Diaz, S. I. (2014b) "El performance de lo yucateco: cocina, tecnología y gusto." *Alteridades* 24(48):59–69.
Bentley, A. (2014), *Inventing Baby Food: Taste, Health, and the Industrialization of the American Diet.* Berkeley: University of California Press.
Everton, M. (1991), *The Modern Maya: A Culture in Transition.* Albuquerque: University of New Mexico Press.
García Gil, G., Oliva Peña, Y., and Ortiz Pech, R. (2012), "Distribución espacial de la marginación urbana en la ciudad de Mérida, Yucatán, México." *Investigaciones Geográficas: Boletín del Instituto de Geografía* 77:89–106.
de Landa, M. (2006), *A New Philosophy of society. Assemblage Theory and Social Complexity.* London: Continuum Press.
Michael, M. (2006), *Technoscience and Everyday Life.* London: Open University Press.
Nakano, Y. (2009), *Where There Are Asians There Are Rice Cookers: How "National" Went Global via Hong Kong.* Hong Kong: Hong Kong University Press.
Ohri, E. (2011), *Long-Distance Nationalism: Constructing "Indian-Ness" in American Kitchens.* Berlin: VDM Verlag.
Rees, J. (2013), *Refrigeration Nation: A History of Ice, Appliances, and Enterprise in America.* Baltimore: Johns Hopkins University Press.
Quintal, E. F. and Quiñones Vega, T. (2013), "Del altar al mercado: los rituales del *pibil* en la Península de Yucatán." Pp. 187–204 in C. Good Esthelman and L. E. Corona de la Peña (eds), *Comida, cultura y modernidad en México: Perspectivas antropológicas e históricas.* Mexico City: ENAH-INAH.
Sutton, D. (2001), *Remembrance of Repasts: An Anthropology of Food and Memory.* Oxford: Berg.
Tuan, Y.- F. (1974), *Topophilia: A Study of Environmental Perception, Attitudes and Values.* Englewood Hills: Prentice-Hall.
Winson, A. (2013), *The Industrial Diet: The Degradation of Food and the Struggle for Healthy Eating.* New York: New York University Press.

7

If you don't use chilies from Oaxaca, is it still *mole negro*?

Shifts in traditional practices, techniques, and ingredients among Oaxacan migrants' cuisine

Ramona L. Pérez, San Diego State University

Introduction

"Coming home" for me is waking up to the smell of bacon and potatoes frying alongside the aroma of freshly brewed coffee and corn *tortillas* being heated on a hot cast-iron *comal* (griddle). I do not know if anyone in our family would feel we were home if my mother changed this traditional breakfast. It is something I think about as I drive the several hours to her house for the weekend and something I have replicated when my own daughters come home to visit. It is also what I long for when I am feeling nostalgic for the comforts of life at a time that held much less stress or somehow just felt safer. The tastes, smells, and particular foods that take us home, to a time and place where we feel engaged with the ones all around us, stand as markers of family, neighborhood, and affection. The memories of and desire for these comforts produce the feelings of nostalgia that grounds this research on migrants from the southern state of Oaxaca and their identification with the cuisine of their points of origin. In this chapter I discuss how the migrants from Oaxaca that have settled in Baja California, Southern California, and Texas have had to change particular ingredients, tastes, techniques of preparation, and practices surrounding the sharing and consumption of their regionally specific cuisine. I begin with a brief overview of migration and the shifting dynamics of family and community in order to ground my research over the last two decades in the changes that have been made to particular dishes, recipes, and practices of production and how these changes have been accepted as the reproduction of "home."

Migration and nostalgia

The mass movement of people across the globe has led to numerous studies on the coping strategies and experiences of displaced persons, with much conversation around strategies for survival, difficulties integrating in the new locale, and the reinvention of community. Much of the research on migratory populations, whether they move from rural to urban spaces within their own countries or they cross national boundaries, demonstrates that for numerous populations, their natal town, with all of its unique stories and images, serves as the new origin story for the generation born in the new locale (Adler 2004; Kondo 2001; Lavie and Swedenburg 2001; Napolitano 2002; Pérez 2012; Van Hear 1998). These feelings and memories provide a foundation for how they reinvent themselves and reproduce community (Akhtar 1999; Burrell 2005; Brown-Rose 2009; Goldring 1996; Gonzalez de la Rocha 1994; Gupta and Ferguson 2009; Hagan 1998; Kim 2010; Levitt and Glick Schiller 2004; Melville 2014; Moran-Taylor 2008; Pérez 2012). Despite the research on the many negotiations migrant populations undergo, one of the more neglected aspects of migration is the powerful emotion of longing or nostalgia for memories of "what once was" before the impetus for migration occurred. I have found that even among those who have no desire to return to their point of origin, there remains a longing for the practices, landscapes, and tastes of home (Pérez et al. 2010). Tied to these many memories are particular dishes that further the unique identification of people to their identities and root them to their point of origin. It is these memories that many try to pass on to subsequent generations through food, cultural practices, and stories so that they too may know the beauty and hardships of life that existed in the place that was once "home."

Over the last two decades, my research has focused on shifts in food practices as they relate to nutrition (Handley et al. 2012; Pérez et al. 2010) and also on how particular dishes are embedded in understandings of taste, texture, and symbolism (Pérez 2014). The research presented here focuses on three migrant communities from Oaxaca who reinvented themselves and created community in San Quintín, Baja California, the city of Arlington, Texas, and the border city of San Diego, California. All three communities continued to define themselves as originating in Oaxaca[1] and considered their unique dishes, ingredients, and customs integral to retaining their identity. The three migrant communities' attempts to maintain their unique cuisine and the challenges they face in doing so, along with their inability to obtain appropriate ingredients, reproduce particular production techniques for a dish due to changes in culinary technology, and to mirror the social and environmental mechanisms that surrounded a particular dish that marked the event as special are the primary focus of the research discussed here.

Oaxacalifornia: Insight on migration between Oaxaca and California

Migration between the southern state of Oaxaca, Mexico, and the United States has been an area of intensive investigation since the late 1980s and was a result of the economic crisis in Mexico and that crisis' profound effect on the countryside (Cornelius and Lewis 2007; Nagengast and Kearney 1990). As Broughton (2008) has argued, migration is not new; rather it has been a significant economic and social strategy for rural communities in Mexico throughout the twentieth century. In fact, Cerrutti and Massey (2001: 187) note that "Migration from Mexico to the United States is the largest sustained flow of immigrants anywhere in the world" and that men dominated these flows until the late twentieth century.

Oaxaca is the largest state in the southeastern region of Mexico, and has some of the highest levels of indigenous language use, poverty, and overall marginalization in Mexico. Its rich history is intimately tied to Mexico's ancient and indigenous legacy and its current populations reflect the diversity of that history. It is home to sixteen different indigenous language families that comprise a wide variety of distinct communities. The languages of these communities have evolved into discrete variants over the centuries, many of which are no longer mutually intelligible despite their lineage to the larger language family. In the 2010 census Oaxaca had 1,165,186 indigenous speakers, representing 34 percent of the total population of 3,801,962. Of these, 83,228 were monolingual in their indigenous language, the majority of which were women (INEGI 2010).

Most migration from Oaxaca is internal to Mexico with the two largest destinations being Mexico City and Baja California (INEGI 2010). The movement of large numbers of Oaxacans to California, many of whom then served as hubs of the migratory network in the early 1990s, and the circular nature of their migratory path between Oaxaca and California, prompted scholars to create the term "Oaxacalifornia" (Kearney 1996; Runsten 1994).

As many Oaxacans stabilized employment opportunities and established families in California and Texas, their status as permanent residents or naturalized citizens also improved. Today, communities with histories of Oaxacan migration can be found with third- and fourth-generation US-born children. Many of these families seek to retain their identity as originating in Oaxaca and to maintain active ties with their home communities, visiting whenever possible and invoking multimedia communication sources in the absence of physical travel. Within Baja California, new communities emerged as a result of migratory networks that are more homogeneous than those in the US, resulting in neighborhoods with families from the same point of origin. The domestic nature of this migration also results in a stronger ability to move people and resources within Mexico that reinforces their ties to home.

Importance of cuisine for Oaxacan populations

The cuisine of Oaxaca is as diverse as its multiple ecological zones that include three major mountain ranges, numerous rivers, coastal jungle, and a wide fertile valley. These ecological niches coupled with local interpretations of foodstuffs that were received through trade, intermarriage, and contact with colonial settlers and missionaries, resulted in a rich cuisine that is not produced in other areas of Mexico. While the basic foodstuffs representative of Mexico are abundant in Oaxaca, the unique tastes that result from Oaxaca's ecological niches and the forms of presentation, preparation, and consumption have produced differences that reflect the unique character of the state and its people. Maize, for instance, is a primary staple in Oaxaca as well as throughout the region, but Oaxacans have mastered the production of a *tortilla* that rivals the size and durability of the flour *tortilla* in northern Mexico. The large corn *tortilla* can be served heated but soft (*blanda*) or toasted (*tlayuda*) to form a sturdy and crunchy base. Both forms of *tortilla* easily span fourteen inches across. The *tlayuda* is served like a *tostada*; it is topped with a paste of caramelized and dry-rendered pork fat, pureed black beans, *quesillo* (a cheese similar to mozzarella in texture that is made from unpasteurized milk), tomatoes, avocado, salsa, and sometimes strips of meat. *Chapulines*, grasshoppers toasted with lime juice and chili, are an important staple food that can be eaten as a snack, sprinkled over a *tlayuda*, folded into a *quesadilla*, or serve as the primary protein in a *taco* (Cohen, Mata Sanchez and Montiel-Ishino 2009). The various *caldos* or clear broths that form the base for many soups have distinct tastes that result from local herbs such as *pitiona* (*Lipia alba*), *yerba santa* (*Piper sanctum*), *chaya* (*cnidoscolus chayamansa*), and *chipíl* or *chipilín* (*Crotalaria longirostrata*). In my research with Oaxacans in Baja California, California, and Texas, *chipíl* was one of the most commonly referenced herbs that they craved and reminisced about in conversations over their hometowns. It is a unique flavor that is used in *tamales* for a vegetarian option or in corn *masa* dumplings that are used in an Oaxacan *caldo* called *sopa de guías* (broth made from young corn, squash blossoms, tendrils from squash plants, and corn dumplings).

Oaxaca is perhaps best known by those outside the region and remembered by those from the region for its seven *moles*. A *mole* is a spicy thick sauce that is served over meat, fowl, or fish and whose recipes can easily include from twenty to forty ingredients. For many Oaxacans the presentation of *mole* at *fiestas* symbolically confirms the importance of the event and the status of the sponsoring family. In many communities it is served only at *fiestas* in honor of the patron saint or other highly revered saint, but not at events that celebrate important events in a person's life such as baptism, marriage, or birthdays. It is expensive and time-consuming to prepare and an inexperienced cook can easily execute it poorly.

The unique tastes of the seven moles of Oaxaca are a result of the diversity of their ingredients and the local chilies, most notably the *chilhuacle* that tastes much like the soil in which it is grown and that is found "only in the dry and semi-tropical mountains of Cuicatlán in the state's La Cañada region" (Culinary Institute of America, Center for the Foods of the Americas). In fact, there are three varieties of *chilhuacle*: *Negro* (black), a rare and extremely expensive dark chile with a very intense flavor; the *rojo* (red) that is used primarily in *mole coloradito*; and the *amarillo* (yellow) that forms the base for *mole amarillo*. The *chilhuacle rojo* and *amarillo* are milder than the *negro* but are more intense than other chiles, except the Oaxacan pasilla of the Mixe region that is smoky and extremely hot. Two other chiles that define the region are the *chilcosle* and *taviche* that are used in *moles* and *salsas* and are toasted and lightly simmered but not stewed, in order to preserve their lighter and fresher taste. The unique taste of chiles from Oaxaca is reflective not only of the region in which they are grown but also in the techniques of preparation that preserve these differences.

Oaxacan cuisine for the twenty-first century

As I have talked with women and their families about their adaptations to their new communities over the last fifteen years, the conversations inevitably turned toward the way in which the foods and food practices in their hometowns could not be replicated in their new location. Rather, the tastes, practices of purchasing and preparing foodstuffs, and the presentation of the meals had to be adjusted or reconfigured in response to local resources. Women reminisced about the relationships they used to have with market vendors and how these networks enabled them to find and use the best quality ingredients. They talked of the huge *fiestas* they would sponsor and the networks of women that would come together to prepare the meals that would feed hundreds. For many women, the ability to establish long-term relationships with market vendors not only assured fair prices and high quality but also served as a mechanism for barter during harsh economic times. The networks of women that became the workforce behind *fiestas* were tied through fictive kinship known as *compadrazgo* and beyond provided labor, also yielded gifts of food and other resources that acted as repayments in this highly sophisticated relationship of exchange. Women's ability to secure such relationships in their home communities defined them as powerful; a scenario that they could not reproduce in the new locale as they moved from domestic labor to paid labor that placed them outside the home for upwards of ten hours each day. While most all of the women I spoke with believed in the necessity of their new work and even felt proud of their ability to move outside the domestic sphere, they also lamented the loss of their social and political role in communal *fiestas*, food preparation, and their ability to provide their daughters with their knowledge of the dishes and practices that define them as a unique people.

One family with whom I spent much time in Arlington, Texas, had mediated some of the strains associated with the different cuts of meat available in Texas as the husband worked as a butcher and could recut meat into the thinner pieces that are preferred in Oaxaca. Despite their ability to control one aspect of their preferred cuisine, however, they were not immune to the shifts in culinary practices that occurred as a result of the wife/mother's long work hours. As a result, the grandfather who worked at the local fried chicken fast-food restaurant provided most of the meals during the workweek. Their children grew up on French fries, fried chicken, coleslaw, and biscuits alongside *tortillas*, *caldos*, beans, and chilies. What the family lamented most was the urban structure of Arlington that precluded the ability to reproduce the religious and cultural celebrations that were centered on the production and consumption of traditional dishes as well as the daily practices that made even basic dishes a reflection of their life. Their children have never known the collaborative process whereby men construct an outdoor temporary kitchen for the dozens of women who would come together to prepare the dishes that would be served over a three-day period of celebration. They do not know how to barter for the kilos of ingredients at the local market that will form the dishes or the taste of *tortillas* made from hand-ground corn on a granite *metate* (milling stone) and heated on a ceramic *comal* over an open flame. They do not know the fusion of clay, granite, wood, and smoke that converts freshly farmed food into the unique tastes of Oaxaca that simply cannot be reproduced in the urban apartments that they now call home. As one of the younger fathers commented to me during a birthday celebration for his wife: "my children will never taste the earth or the rain in their food. For them, food comes in boxes or wrapped in plastic. It makes me sad to think that their blood will weaken and one day carry no memory of our home."

Whether women worked in the agricultural fields of Baja California, in dry-cleaning stores or fast-food restaurants in Texas, or as childcare providers, maids, or factory workers in California, most of them had come to rely on either elderly family members or older children to prepare the daily meals for the family. In fact, in 2009, while working with the community of San Ramon in Baja California, a household survey demonstrated that in sixty-eight of the seventy-two households surveyed, an older child prepared the meals during the workweek. The need to incorporate modern technologies—such as microwaves, blenders, toaster ovens, and food processors—that eased the food-preparation burden for younger family members, along with the use of prepared foods required that traditional modes of preparing and eating meals be modified to accommodate their new lifestyles. For instance, the preparation of *salsa* in a granite *molcajete* (mortar) allows for greater control over the texture and taste of each ingredient. A skilled cook can forcefully mash garlic against the sides of the *molcajete* so that it becomes part of the liquid while onions, tomatoes, tomatillos, and chiles can be left in various stages of breakdown that allow for a burst of sweet or sour tastes, or heat in the mouth. It is not only the quantity of each of these ingredients that can make a difference in taste but also the actual techniques of preparation that enhance or respect the flavor and texture of each ingredient. The use of the blender precludes this process and as a result *salsas* become less distinct and their use as a special condiment with different emphases of taste and texture that can be paired with a particular dish is lost. Instead, *salsa* becomes a standard condiment of tomato, onion, garlic, and chile that is used as a flavor enhancer similar to salt or lime.

Many families talked of the loss of cultural experience surrounding food and the loss of knowledge associated with the preparation of particular dishes, such as the *salsas* noted above, that unfold through the chopping, sautéing, simmering, and incorporation of ingredients in sequences that extract the flavors of each ingredient. Interestingly, while the microwave has become a standard kitchen tool for the warming of prepared foodstuffs, it is not used to heat *tortillas* as doing so greatly modifies the texture of both flour and corn *tortillas* and renders them tasteless. While rarely prepared over a ceramic *comal*, this is one element of traditional cuisine that is stilled prepared over a *comal* albeit one of cast iron or aluminum. Alongside these shifts in production, technique, and consumption are the challenges in acquiring the ingredients that define a dish.

The issue of sourcing the specialty ingredients that define displaced persons' particular dishes as unique from those of neighboring regions has been an issue throughout time. Although technological advances have allowed many ingredients to be shipped across the globe, the freshness and unique qualities of locally produced ingredients is rarely obtainable. In their place are similar but mass-produced items that are part of an export process and that come to represent a wider depiction of a culture group. For instance, while *tamales* are prepared throughout Mexico and have been turned into a pan-representation of Mexican cuisine, their taste and style can vary dramatically: for example, the small, thin *tamales* that contain a small amount of meat and which are wrapped in dried corn husks are reminiscent of the north of Mexico; the long, wide *tamales* that are stuffed with eggs, olives, and meat and wrapped in banana leaves are found on the Caribbean side of the country. Recipes may be similar but the taste of the ingredients differs as a result of soil, water, seed, light exposure, and much more. In the US, the diversity of foodstuffs that are received from across the globe is perceived as more than adequate for the reproduction of most ethnic foods. Yet the reality is that fresh ingredients such as herbs, chiles, tomatoes, cheese, and many other staples of the Oaxacan diet, simply cannot be replicated, and this adds to the burden of changing modes of preparation and presentation in reproducing "home." A good example of this difference occurred while I was working with a group of Mixtec women during a *fiesta* in San Ramon in Baja California.

As I was documenting the recipe and techniques for preparing *mole negro*, I noticed that they were not using the *chilhuacle* or Oaxacan *pasilla*. In fact, as they were putting the finishing touches on the *mole*, I asked them why they had chosen the particular chilies that they had used. Lorena, the woman in whose house we were, answered with a puzzled look: "Pues, porque son los unicos disponibles acá" ("well, because they are the only ones available here"). Hesitantly, I responded that the *mole* wasn't actually as dark as I was accustomed to eating in Oaxaca and that the taste was a bit different. In fact, although I didn't say this, the *mole* tasted more like my grandmother's *mole*, and she was from Jalisco. We then entered into a conversation about how the food tasted differently in Baja California but that they had grown accustomed to it. I finally asked, "pues, entonces si no usa chilies de Oaxaca, ¿todavía es Mole Negro de Oaxaca?" ("so, if you are not using chilies from Oaxaca, can it still be Oaxacan *mole negro*?") The women paused for a moment and looked around at each other before bursting into laughter. "Si," Lorena answered, "estuvo preparado por Oaxaqueñas!" ("yes, it was prepared by Oaxacans"). As the laughter died down, she went on to say that they would never serve this version to visitors from Oaxaca—or would at least do so only with great embarrassment (*pena*). Rather, once a month a bus from Oaxaca comes in with various chilies, spices, herbs, vegetables, and fruit along with ceramic vessels, *petates,* and other customary items that are produced for local use in Oaxaca. The items are expensive, she said, but worth it as they bring home back into their lives, if only for a brief moment.

In the US, some communities are lucky enough to have transnational care packages that carry specialty food and other items between them and their home communities in Oaxaca through a system called *envios* (literally, exported goods). Under this system, a local person who can move legally between Oaxaca and the US transports packages of locally grown and produced foodstuffs from migrants' communities in Oaxaca to their new communities in the US (Grieshop 2006; Pérez 2010). While somewhat costly, the opportunity to receive products from home allows migrant communities to reinvent the tastes and, in some cases, the practices from their home communities. For now, the techniques associated with many technologies of food production, such as the *molcajete* and the *comal*, are still reproduced in multigenerational families. As family members who grew up in Oaxaca are lost or directed away from food production, however, the techniques that produce the unique tastes and textures associated with these technologies may not be reproduced and thus will not be preserved.

Compromises

Nostalgia is defined as a sentimental longing or wistful affection for the past; a wistful desire to return in thought or in fact to a former time in one's life, to one's home or homeland; the state of being homesick; a bittersweet longing for things, persons, or situations of the past. These and similar definitions abound, but the definition that I find the most relevant to the way in which migrants from Oaxaca talk about, attempt to reproduce, and seek to provide for their children who do not know their home communities, comes from Merriam Webster's online dictionary:

> pleasure and sadness that is caused by remembering something from the past and wishing that you could experience it again.

Food—its taste, smell, production, and presentation—has perhaps the biggest association with nostalgic feelings for home. The changes in these areas as a result of displacement, whether chosen or enforced, has not been a significant area of research in studies on migration and resettlement and yet they should be. The ability to taste home, to reproduce the warmth of a kitchen fire or the closeness of women's collaborative food preparation, to remember the laughter, conversations, and intimacy of a family meal, to celebrate life's greatest moments with the tastes, smells, and textures of home—these are the memories that can give us comfort when we most need it and can help us transition from one place to another with a greater sense of safety and comfort.

Note

1 Although I use the term "Oaxacan" in this chapter, it should be noted that one group was formed from mestizos, another formed around two indigenous groups from the Mixteca region, and another was a combination thereof.

References

Adler, R. H. (2004), *Yucatecans in Dallas: Breaching the Border, Bridging the Distance*. Boston: Pearson.

Akhtar, S. (1999), "The Immigrant, the Exile, and the Experience of Nostalgia." *Journal of Applied Psychoanalytic Studies* 1(2):123–130.

Burrell, J. L. (2005), "Migration and the Transnationalization of Fiesta Customs in Todos Santos Cuchumatán, Guatemala." *Latin American Perspectives* 32(5):12–32.

Broughton, C. (2008), "Migration as Engendered Practice: Mexican Men, Masculinity and Northward Migration." *Gender and Society* 22(5):568–589.

Brown-Rose, J.A. (2009), *Critical Nostalgia and Caribbean Migration*. Volume 23 of Caribbean Studies. New York: Peter Lang Publishing.

Cerrutti, M. and Massey, D. S. (2001), "On the auspices of female migration from Mexico to the United States." *Demography* 38(2):187–200.

Cohen, J. H., Mata Sanchez, N. D., and Montiel-Ishino, F. (2009), "*Chapulines* and Food Choices in Rural Mexico." *Gastronomica: The Journal of Critical Food Studies* 9(1):61–65.

Cornelius, W. A. and Lewis, J. (2007), *Impacts of Border Enforcement on Mexican Migration: The View from Sending Communities*. La Jolla, CA: Center for Comparative Immigration.

Culinary Institute of America. Center for Foods of the Americas: Oaxaca. http://www.ciaprochef.com/CFA/mexico/oaxaca.html. Accessed January 12, 2015.

Goldring, L. (1996), "Gendered Memory: Constructions of Rurality Among Mexican Transnational Migrants." Pp. 303–329 in E. M. DuPuis and P. Vandergeest (eds), *Creating the Countryside: The Politics of Rural and Environmental Discourse*. Philadelphia: Temple University Press.

Gonzalez de la Rocha, M. (1994) *The Resources of Poverty: Women and Survival in a Mexican City*. Oxford, UK: Blackwell.

Grieshop, J. I. (2006) "The Envios of San Pablo Huixtepec, Oaxaca: Food, Home and Transnationalism." *Human Organization* 65(4):400–406.

Gupta, A. and Ferguson, J. (2009), "Beyond 'Culture': Space, Identity and the Politics of Difference." *Cultural Anthropology* 7(1):6–23.

Hagan, J. M. (1998), "Social Networks, Gender, and Immigrant Incorporation: Resources and Constraints." *American Sociological Review* 63:55–67.

Handley, M. A., Robles, M., Stanford, E., Collins, N., Seligman, H., Defries, T., Perez, R. L., and Grieship, J. (2012), "Navigating Changing Food Environments – Transnational Perspectives on Dietary Behaviours and Implications for Nutrition Counseling". *Global Public Health: An International Journal for Research, Policy, and Practice* 8(3):245–257.

INEGI (Instituto Nacional de Estadistica y Geografia) (2010), *Conteo de Poblacion y Vivienda 2010. Cultura.* 12 de junio 2010. www3.inegi.org.mx. Accessed September 12, 2014.

Kearney, M. (1996), *Reconceptualizing the Peasantry: Anthropology in a Global Perspective.* Boulder, CO: Westview Press.

Kim, W-B. (2010), "Nostalgia, Anxiety and Hope: Migration and Ethnic Identity of Chosŏnjok in China." *Pacific Affairs* 83(1):95–114.

Kondo, D. (2001), "The Narrative Production of 'Home,' Community and Political Identity in Asian American Theatre." Pp. 97–117 in S. Lavie and T. Swedemburg (eds), *Displacement, Diaspora, and Geographies of Identity.* Durham, NC: Duke University Press.

Lavie, S. and Swedenburg, T. (2001), *Displacement, Diaspora, and Geographies of Identity.* Durham: Duke University Press.

Levitt, P. and Glick Schiller, N. (2004), "Conceptualizing Simultaneity: A Transnational Social Field Perspective on Society." *International Migration Review* 38(3):1002–1039.

Melville, G. (2014), "Identity Strategies and Consciousness Shifts of Sanmiguelense Mixtec Youth in Transnational and Transcultural Spaces." *Latin American Perspectives* 41(3):178–193.

Merriam Webster Online Dictionary. http://www.merriamwebster.com/dictionary/nostalgia. Accessed January 25, 2015.

Moran-Taylor, M. J. (2008), "When Mothers and Fathers Migrate North: Caretakers, Children, and Child-Rearing in Guatemala." *Latin American Perspectives* 35(4):79–95.

Nagengast, C. and Kearney, M. (1990), "Mixtec Ethnicity, Social Identity, Political Consciousness and Political Activism." *Latin American Research Review* 25(2):61–91.

Napolitano, V. (2002), *Migration, Mujercitas, and Medicine Men: Living in Urban Mexico.* Berkeley: University of California Press.

Pérez, R. L. (2012), "Crossing the Border from Boyhood to Manhood: Male Youth Experiences of Crossing, Loss and Structural Violence as Unaccompanied Minors at the US–Mexico Border." *International Journal of Adolescence and Youth* 19(4):1–17.

Pérez, R. L. (2014), "Tecno-estética translocal y alimento en MexAmérica." Pp. 299–316 in S. I. Ayora-Diaz and G. Vargas-Cetina (eds), *Estética y Poder en la ciencia y la tecnología: Acercamientos Multidisciplinarios.* Mérida: Ediciones de la Universidad Autónoma de Yucatán.

Pérez, R. L., Handley, M., and Grieshop, J. I. (2010), "Savoring the Taste of Home: The Pervasiveness of Lead Poisoning From Ceramic and its Implications in Transnational Care Packages." Special Issue, *Anthropological*

Perspectives on Migration and Health. Craig Hadley (ed.), *Annals of Anthropological Practice* 34(1):105–125.

Runsten, D. G. (1994), *A Survey of Oaxacan Village Networks in California Agriculture*. California Institute of Rural Studies.

Van Hear, N. (1998), *The Mass Exodus, Dispersal, and Regrouping of Migrant Communities*. London: University College Press.

8

Changing cooking styles and challenging cooks in Brazilian kitchens

Jane Fajans, Cornell University

Brazilian cuisine has recently emerged as an exciting new addition to the global food scene. This appearance results from the creative activities of a number of top chefs and parallels the changing foodscape across much of Latin America. These chefs have introduced a cuisine that incorporates intrinsically local Brazilian ingredients into dishes that emerge from a new manifestation of nouvelle cuisine. These dishes and these restaurants illustrate a creative merging of continental techniques with ingredients that embody regional, ethnic, and familial associations. The development of this new culinary practice has garnered renown in the gastronomic community, and several of these chefs and restaurants have ranked among the top artists and venues in the world.

The celebration of top chefs and their creative uses of traditional and global food techniques and ingredients have had significant influences on attitudes and food practices outside of these choice restaurants. This chapter examines significant changes in the way people are now eating at home. Although few Brazilians are attempting to cook haute cuisine in their homes, many are now embarked on an exploration of new foods and new techniques. This path changes who cooks for whom and how and where such cooking is done. In contrast with the food preparations and activities that did and often still do feed the majority of the national population, a growing number of middle- and upper-class Brazilians are spending time and money in the kitchen.

Wealthier families have long appreciated fine dining. They have frequented the many continental (mostly French and Italian) restaurants in the big cities, and served dishes such as stroganoff and steak with béarnaise sauce to company in their homes. In an increasingly affluent society, however, kitchens have become a focus of consumption and display. As food and gastronomy are accorded greater importance, more households are upgrading their kitchens and renovating to make the kitchen a more central part of the social life of the household.

Who cooks and for whom?

Until recently, the middle and upper classes did not cook for themselves but employed domestic labor to shop, cook, clean and serve the family. The number of such workers in a household varied according to the status and size of family but could often equal or exceed the number of family members, as staff specialized in cooking, childcare, cleaning, gardening, and chauffeuring moved into and out of the space. In such a household, the physical space of the house or apartment tended to be divided between that of the family and that of the domestic workers. These two spaces were/are called the social space and the service space. Each area would usually have a separate entrance, although they would be internally connected by one or several doors. As lifestyles have changed and international influences become better known through TV and other media, so too has the use of space. Relegating the kitchen to the service space, and maintaining the social space as intrinsically formal, has given way to a more informal style and greater flow between the areas.

Traditional Brazilian kitchens were located in the service area of an apartment or house. They were generally separated by walls and closed doors that were opened only when meals were being served, frequently by the servant who prepared the meal but sometimes by additional domestic workers. In lower- or middle-class homes, the cook might be the same person who shopped, cleaned, and looked after the laundry, etc, but in wealthier or more elite households several people might fill these different roles. The kitchen was not a central living space as it is in many traditional cultures, but rather an appendage to the laundry and maids' rooms.

Outfitted with a propane stove, cold water, and a small refrigerator, the kitchen fixtures were merely basic. Hot water was often not available on tap, but cooks could boil it for drinking and cleaning. Since there were servants to shop daily, storage was minimal. Equipment consisted of several pots and pans of various sizes, basic utensils, and very few small appliances. During my first stay in Brazil in 1982 the two pieces of more specialized equipment that most cooks deemed essential were a pressure cooker and blender. The staple components of the traditional meal, rice and beans, usually need only pots, a heat source, and water to cook, but they cook faster and more economically if the household possesses a pressure cooker. So a pressure cooker was one of the first pieces of equipment to be added to the basic repertoire. The second piece of equipment was frequently a blender that can be used to puree beans, or make fruit smoothies, often for alcoholic drinks among other things. A filter to purify water was also important, as it avoided the necessity of boiling a large pot and cooling it for drinking water. Some households had charcoal filters attached to the sink while others use a ceramic filtering urn. Nowadays, many households purchase water, often in large multi-gallon bottles set into dispensers.

The quotidian meal

In this simple space, simple meals were/are prepared. Clichés usually contain a kernel of truth, and it is true that the Brazilian diet relies heavily on rice and beans. This ideal meal is for many, however, just that, an ideal. Many Brazilians do not consume these elements as often as they give lip service to their partnership. Some proportion of the population can afford these foods only occasionally (Goldstein 2003; Scheper-Hughes 1992); others see them as the lowest common denominator of a meal, and often consume more prestigious items. However, these foods retain a strong affective value. In talking to expatriate Brazilians I am not surprised to hear that many who did not eat rice and beans regularly in Brazil now find that they are cooking them frequently with a sense of nostalgia, but also a sense of ingesting an important component of their identity as Brazilians. If we are what we eat, then eating the standard Brazilian diet keeps people Brazilian.

The combination of rice and beans has evolved over time and across many countries (Abala 2007; Wilk and Barbosa 2012). The two foodstuffs have been called the perfect marriage and the complete meal because between them these two starches provide a full amino acid component (Barbosa 2012; Embrapa 2009). Adding a piece of meat, or flavoring the beans with meat or bones, provides additional nutrition and a good deal of flavor. Until recently, the typical meal consisted of rice and black beans accompanied by a piece of meat when available, and in most middle- and lower-class households, rice and beans remains the staple core although other starches and a small amount of meat provide variety. Most meals would be considered lacking without the inclusion of toasted manioc meal, *farinha*, which gives additional texture to both the main dish and other elements on the plate. In all these ways the ideal meal is more than the sum of its parts. Because of the reliance on hired cooks, even among middle-class households, shopping and cooking has mostly been provided by domestic help, who are usually members of the poorer social class. Cooking what they know, they often cook only a slightly upscale variation of the traditional foods they are familiar with. In addition to the ubiquitous rice and beans, the better-off households consume larger and more varied portions of meat and a greater array of fresh vegetables.

The hegemonic form of this typical meal is the bean and meat casserole called *feijoada*. *Feijoada* is frequently referred to as the national dish of Brazil. It is traditionally served on Saturday afternoons and typically constitutes an occasion to gather a large extended family together. The assumption that everyone eats this meal at the same time across the nation is more myth than reality (see Fajans 2012) but the belief serves as a unifying ideal in a diverse nation. If everyone really did eat *feijoada* at the same time, then they would share a common essence across families, race, and class, but as mentioned earlier this is not the case. *Feijoada* is a bean stew that contains a variety of meats. An authentic *feijoada* ought to include the "low" pieces of a pig: the trotters, hocks, tail, and ears are traditional components (Fry 2001). To this base a variety of other meats are frequently added, including beef tongue, dried beef, *linguiça* (pork sausage), and some smoked meat (often the hock or tail). *Feijoada* is the main dish, but a *feijoada completa* is a whole meal and includes a fixed array of complementary dishes: rice; *couve* (sautéed collard greens), orange slices to squeeze over the beans; a *salsa* made of raw tomato, peppers, hot peppers, and cilantro; and toasted manioc flour often mixed with egg and onion, *farofa*.

The preparation of this massive meal is usually performed by the household help mentioned above, although I have eaten *feijoadas* cooked by the family themselves when they are in their country homes. The fact that this meal requires a good deal of labor to prepare, serve, and clean up, means that domestic workers are required to give up days they might otherwise expect to spend with their own families. It also belies the myth that all families sit down together to enjoy this repast. The cooks and servers may enjoy some of the meal's leftovers and even take some home to their families, but they do not share the commensal experience often attributed to this tradition. The class divide continues even if all classes eat the same food; the quantity of food one eats can also be a marker of status even if everyone consumes the same dish (Goody 1982).

Although rice and beans are found consistently throughout the country, they are often served in particular styles, and the type of bean used varies regionally. *Feijoada* made with black beans is associated with Rio de Janeiro and São Paulo. People from Minas Gerais prefer brown beans. When the dish is made with red beans, it is classified as Bahian. When people in the northeast of Brazil cook beans, they tend to favor black-eyed peas and they tend to eat their peas with clarified butter. These regional variations have become dispersed throughout the country, however, as migration to the big cities draws people from all regions for employment, often as domestic workers, in other parts of the country. Although eaten in Rio, my first impressions of

Brazilian food were produced by a Mineran cook and those dishes became my foundational recipes.

In contrast to the foods traveling with the cook or maid from one region of Brazil to another, in a number of communities, domestic cooks have been taught ethnic and national cuisines and techniques brought by their employers to Brazil. In these households, recipes may be built around the cuisines of Germany, Ukraine, Japan, Lebanon, certain African countries, and other immigrant homelands. Often these cuisines are not just taught to one cook or maid but subsequently passed down through several generations as these jobs are frequently also passed down through a single family's generations. In such households, meals may be more varied than those of the lower and rural populations and even those of the upper classes. These households may eat a diet only slightly varied from that of the homeland. Kitchens in these houses may be outfitted with special equipment required for making pasta, dumplings, cakes, and custards, or sushi, stir-fries, and baklava.

Traditionally the main meal of the day was the midday meal. This pattern is common throughout rural communities but continued until recently in urban areas as well. A large midday repast satisfied those who had been working since early morning. It also enabled domestic help who were not living on the premises to shop, prepare, and clean up a family meal and leave at the end of the afternoon. Such workers often left a light meal to be heated up or eaten cold in the evening as the household desired. As more adults now work outside the home and often have to travel significant distances to their work places, the tradition of the big midday meal at home has disappeared in urban areas except on weekends. As such lifestyle changes have occurred, so has the pattern of eating. Dinner has become the main occasion for a family to gather together after a day apart. This meal has also gained in importance and prestige. This increase in status derives in large part because upper-class Brazilians have been able to mix meals produced by domestic help at home with those procured by dining out. The wealthy can afford to eat well away from home and eating out is a prestigious activity.

Food has always been available outside the home, and street foods, beach stalls, and market eateries dot the landscape offering everyday and local fare. To a great extent, what people eat away from home mimics the food they expect to eat at home. The most common dish served at small eateries, street stalls, bus stations, etc is what Brazilians call the *prato feito* or ready meal. It too is built around the dyad of rice and beans but it includes a small piece of meat and, often, another starch such as bread or pasta, as well as a small salad or slice of tomato. This meal can be found in cafeterias, street stalls, or bus and train stations. It can also be individually created by customers in the many *a quilo* restaurants (by the kilo/*quilo*), buffet-style restaurants that sell food by weight. These restaurants are now ubiquitous in all urban areas, including university dining halls, office buildings, shopping centers, and street-food malls. Despite the variety of venues, the offerings hold the promise of a satisfying meal for a variety of different customers, offering several kinds of beans, rice cooked in a number of ways, a selection of meats (often some cooked on a grill), and hot and cold vegetables. Since the price is by weight, patrons can choose how much they want to eat or how much they want to pay. In contrast to the buffet-style restaurants that sell familiar food by weight, a wide assortment of fast-food restaurants constitute alternative venues for inexpensive meals outside the home. Brazil has both transnational fast-food restaurants like McDonalds and Pizza Hut, and local spin-offs like Giraffe. These restaurants are readily accessible to most urban restaurants but are not as common in the more rural areas.

Fine dining and the new Brazilian cuisine

In contrast to these omnipresent eateries, "fine dining" for many years meant Continental, French, or Italian restaurants. Today, though, it has increasingly come to include the new Latin American Cuisine and Asian and fusion styles as well. Restaurants always represented a realm where good taste of various sorts could be acquired and displayed. Now, however, these attributes can be acquired in numerous venues and more and more resources are popping up to support such status. The biggest change is the creation of a new Brazilian cuisine, one which derives in large part from the new haute cuisine of Europe and other Latin American countries. In these venues, ingredients from the more distant regions of Brazil such as the Amazon, the Mata Atlantica, the Cerrado, and the Pampas, to name just a few, are playfully incorporated into the dishes of chefs trained in the European and international arena. Ingredients from their own country, previously unknown to urban audiences, are becoming familiar and available to even the home cook (at a price). The continuum between national and international, home and restaurant cooking is more blurred than before.

Transforming the value of cooking

In concurrence with this new food interest, middle- and upper-class families have begun to explore cooking and entertaining in their own homes. Gourmet food shops are increasingly visible, bookstores have expanded sections of local, regional, and international cookbooks, and fancy appliances are advertised everywhere. Cooking classes are springing up in cosmopolitan areas, and people are planning travel around culinary interests. To facilitate these new interests, wealthier families are starting to renovate their houses and apartments to include upgraded kitchens. Significantly, these new kitchens are being incorporated into the living areas of the house or apartment instead of being separated from the 'social' spaces. Walls are being removed or opened up, seating is being added to the kitchen or 'breakfast bar,' and high-tech appliances like chef's stoves, top-end refrigerators, dishwashers, microwaves, espresso machines, and designer cookware are being advertised and purchased. Cookbook stores with exhibition areas for cooking demonstrations are opening up, cooking shows are omnipresent on TV, and restaurant chefs are national celebrities.

The investment in these new kitchens is the greatest indicator that cooking has become a fashionable endeavor. Given the underlying class structure in Brazil, such renovations would not be undertaken just to assist a domestic worker. The new spaces are for display as well as function. The shift in emphasis is interesting; previously adults left their homes to mingle in public and accrue status through external factors such as the food they ate or the clubs they frequented. Domestic workers or housewives did the cooking. Nowadays fashionable couples are inviting people into their homes to demonstrate their personal cooking and entertaining skills, and thus acquire a different kind of renown. Cooking in the home is no longer an exclusively female occupation (restaurant chefs were frequently males but that was an occupation outside of the house). Couples are increasingly investing their time and money to increase their culinary capital. A recent article in the Brasilia magazine, *Comer no Brasilia* (Eating in Brasilia) (2014), describes how men, women, and couples are venturing into the world of gastronomy. Travel may now focus on enrolling in cooking classes abroad and at home and following gastronomic trails. Some enthusiasts are even hiring well-known chefs to come to their houses to offer private cooking lessons. The cachet of learning the signature dish from a popular restaurant propels these enthusiasts to try new culinary skills.

One can display a number of different types of status through the medium of food. People can display their cooking skills and recipes but also their *material savoir faire* through china, silver, table ornaments, collections of cookbooks, equipment, and relics from regional and international travel. In addition they can display certificates, photos, and affidavits from gastronomic professionals that attest to their knowledge and network of connections.

Encompassing the nation in the dining room and kitchen

Until recently, European-style and dishes held precedence, and ingredients and serving styles reflected the greater prestige of European forms, especially in middle- and upper-class homes. These homes and meals frequently mimicked the food found in restaurants. Food was served on platters and often passed around by domestic workers. As noted above, tableware included china, crystal, and silver pieces, and ornaments consisted of candles and flowers. Now, however, greater regard is given to domestic, regional, and national specialties and the local milieu in which they might be prepared and consumed. Along with this renewed appreciation of the local and national, people are reevaluating local, often folkloric, styles of décor. Brazil has a national institute called the Institute of National Historic and Artistic Patrimony (*Instituto do Patrimônio Histórico e Artístico Nacional*), commonly known in Portuguese as IPHAN, which has been charged with selecting the best of Brazil's historic, cultural, and artistic treasures. What is most interesting is that IPHAN has been charged not only with identifying paintings, buildings, palaces, and churches, but also with selecting those less grandiose treasures that are central to the conception of Brazilian culture. An appreciation of local culinary dishes is accompanied by local pottery, glassware, tableware, and drinks. As Brazil seeks to identify, catalogue, and protect the unique traditions of its regional and ethnic diversity, these icons of identity acquire greater value. Certain styles of cooking, manner of dress, venue for consumption, and forms of presentation acquire new value. This array of tangible and intangible patrimony plays into contemporary ideas of status and sophistication (http://flavorsofbrazil.blogspot.com/2010/04/utensils-clay-cookware-from-goabeiras.html).

As a large country spanning numerous geographical and climatic environments, Brazil's different regions have distinctive styles of cooking using local ingredients and cultural traditions. These too have been granted patrimonial value that is being registered with IPHAN. Alongside the different foods, recipes, and styles of serving, many of the regions have evolved or preserved different dishes for serving local and ethnic specialties. Clay dishes are found throughout Brazil but they have particular characteristics. In Espirito Santo, the clay pots (*Capixaba* pots) are made from dark-grey river clay and molded by hand, building base and sides with snakes of clay pressed together and smoothed with a polished stone. After drying, they are baked in an outdoor fire, not a kiln. While they are still hot, they are beaten with the root of a river mangrove plant that exudes a black sap containing tannin. This sap is applied while the dish is still hot. It coats the fired clay and imparts a waterproof finish to it. These pots are round with a flat bottom and thick sides that curve slightly inward at the top to form a rim on which a lid made of the same clay can be rested. Important to the cultural patrimony of this craft, these pots are made in small community collectives often run by extended family groups. These groups have a monopoly on the production of such pots. They market their wares at the rustic workshops or in outdoor markets and a few craft stores. Most of the restaurants in town serve stews, soups, and rice and beans in these pots. The uniqueness of these pots and the manner of their production in extended family cooperatives has gained both the pots and their makers a form of notoriety. They are now being considered for a listing in the national patrimony.

In contrast to the pots made in Espirito Santo, the clay dishes made and used in Bahia are made of red clay. These dishes are much thinner and take the form of a wok with a curved bottom and outward flaring sides. They do not have a lid. Many are made on a rustic potter's wheel and glazed with a red slip on the inside. Lighter and more delicate than the *Capixaba* pots, they are not, however, granted the same cultural value. Both pots are used on the top of the stove where their contents are boiled and simmered. What is cooked in each of the pots, however, is seen as part of the intangible heritage of each region and both types of pots are used for specific meals, principally a type of seafood stew called *moqueca*.

Although both regions cook a style of *moqueca*, the ingredients used in each state are different and have a different heritage. The Bahian *moqueca* uses ingredients from Europe, Brazil, and Africa, and is valued at least in part because of this blend that parallels the heritage of those who prepare it. The *Capixaba moqueca* does not symbolize the same ethnic mélange, and its ingredients are not seen to have the same associations, although this lack of association may be viewed by the community as indicative of its purer origin. There is considerable rivalry between the Bahians and the Capixaba over who makes the best and/ or most authentic *moqueca*.

A third type of regional cookware is made in the state of Minas Gerais. These pots are carved out of soapstone and have a copper band around the rim to which handles are affixed. The lids are also made of stone. These pots are made in specialty factories and are very heavy. The stone material and weight means that the pots are excellent for slow-cooking stews and casseroles. They can be used on top of the stove or in gas, electric, or wood-fired ovens. They are slow to heat but also slow to lose heat. Food can be served straight from these beautiful implements.

In the northeast of Brazil, *farinha* is often served in a special dish or bowl that is carved in the shape of an animal such as a duck or anteater. The tail of the animal is carved as a spoon and is used for serving the manioc flour. This dish sits on the table through all the meals and becomes an important fixture of the table setting. People serve themselves *farinha* from these bowls as we might sprinkle sugar on our cereal.

Several regions also use gourds as specialized food and drink containers. In the south of Brazil, especially bordering Argentina and Uruguay, people drink an infusion made from the plant *yerba mate*. This drink, *mate* or *chimarrão*, is traditionally drunk from a gourd cup through a metal straw that filters out the pulp. A cup or gourd of *mate* is frequently passed communally around a group who add more hot water when the level drops. At the other end of Brazil, in the Amazon region of the north, a regional dish called *tacacá* is traditionally served in a bowl made from half a large circular gourd. This semi-spherical bowl is frequently incised with geometrical designs and may be set in a woven basket to keep it from tipping. *Tacaca* is essentially a soup made from boiled manioc juice (squeezed from the grated tuber), manioc starch, dried shrimp, and a leaf called *jambu* which produces a sharp tingling sensation, almost like a bolt of electricity in the mouth.

These are just a few examples of the types of cooking equipment that one can find across the country that evoke the different regions, populations, recipes, and special ingredients. As restaurant chefs become increasingly interested in drawing from the rich resources of the Brazilian nation, they are also highlighting the crafts and values of these regions. This type of display inspires others to travel to taste the regions and bring home some of the specialties found around the country. Contemporary kitchens may incorporate displays of various equipment and artifacts in imitation of renowned restaurants like *Brasil a Gosta*, or *D.O.M.* in São Paulo, which have recently redecorated their interiors to showcase dishes, plants, and artifacts from around the country.

Media matters

The restaurant has long been a model of cultural and fashionable eating, but media are an increasingly important influence on what people eat, how they prepare it, and where they prepare it. Having tried a variety of foods in restaurants, the newly inspired cooks seek to replicate them at home. To do this they browse magazines that offer increasingly exotic recipes; consult cookbooks; watch TV cookery shows; and search online from sources like YouTube. These electronic devices are the most recent kitchen equipment to be added to the repertoire.

Brazil, like many other cultures, has a rapidly expanding media culture. When I first started working in Brazil I noted many magazines in the home-decorating genre. Kitchen and bathroom remodels seemed to titillate middle- and upper-class women (even if they did not succeed in making many such transformations at that time). This interest in kitchen remodeling was followed by an influx of cookbooks, of both domestic and international varieties. As a multicultural nation, Brazilian food has long included recipes from Portugal (the "motherland"), Italy, Germany, Japan, Angola, Nigeria, Lebanon, and many other African nations. More recent immigrant communities include Chinese, Ukrainian, Polish, and those from other Latin American countries. As noted, cookbooks including cuisines from these and many more cultures can be found in bookstores across the country. In addition, specialty cookbooks on different kinds of diets, cakes and desserts, soups, and wine and beer have proliferated. The size of the cookbook section has grown enormously in the major bookstore chains in the country. Many of these lush and beautifully illustrated cookbooks adorn coffee tables and display shelves in people's homes. Last, but certainly not least, cooking shows have become a popular form of entertainment. There are standalone cooking shows that feature particular chefs; cooking demonstrations as segments during daily TV shows; and interviews with celebrity chefs and tours of restaurants. In addition, many individuals make and upload their own cooking demonstrations featuring regional specialties such as *moqueca, pato no tucupi*, or *feijoada*, but also some family favorites such as *brigadeiros* (fudge balls rolled in sprinkles or sugar) or wedding cookies.

With all of these resources to help people broaden their experiences and gain competence in cooking, gastronomy has entered mainstream consciousness. In connection with this, the knowledge of ingredients available to interested persons has proliferated. I have written elsewhere about my difficulties finding regional cookbooks with local recipes when I first went to Brazil (Fajans 2012). This lacuna was accompanied by an inability to get many of the special ingredients required for these recipes outside of the region in which they are traditionally served. Both of these shortcomings have vanished in the last several decades for several reasons. As mentioned above, middle- and upper-class people have become more interested in a wide variety of foods. Along with this, there is greater mobility in the country. People travel to regions as tourists and include culinary exploration as part of their plans. In addition, more and more people are moving and migrating to regions far from those they grew up in. Many of these people seek to reproduce the favorite or comfort foods of home and have created a market for such products. Regional restaurants have also sprung up in far-flung cities. These restaurants serve two purposes: the first is to provide familiar food to such migrants, and eating places serving regional foods often serve as meeting places or social clubs for them; the second stems from a kind of entrepreneurship in which people realize they can make a living providing food to local or inquisitive people. There are still some dishes and ingredients that do not "travel" and are more rooted in their traditional geographic space. This may be because they are more easily perishable or because there is still not a big enough market. Such ingredients are often hand carried by friends and relatives to those far away. Most of the people I queried about "home" food had a particular dish they craved and which they wanted as their first dish back at home. Many residents from a region longed for the same two or three dishes indicating that these dishes had become a kind of icon of home, and a marker of identity.

As mentioned above, another major change in the country is the proliferation of cooking classes, schools, and programs. When I began my research, I applied for a grant to enable me to study cooking at a cooking school in Brazil to learn techniques and philosophy about meal planning, preparation, and presentation. Going online to search such programs, I found only two, and talking to people in tourist bureaus brought another to light. By the time I received my money, at least one of these programs was no longer functioning; one was teetering, and the third turned out to be a professional training program that would not accept amateurs like me. The one place that did give cooking lessons was the language school I attended (as do many such programs geared for tourists). This course was intended for non-Brazilians and was aimed to introduce the foreigners to regional dishes. The school offered three extracurricular classes: cooking, samba, and capoeira. These were often cited as the three most distinctive aspects of the Bahian culture (the state in which the language school was located). These classes were oriented towards acquiring something classed as exotic and foreign, and not as expanding everyday knowledge.

Nowadays, however, gastrotourism, cooking classes, and food fairs have become much more mainstream, and people know about them even if they do not themselves participate. Such food-focused events provide entertainment, social engagement, and increasingly social/culinary capital (Naccarato and Lebesco 2012). Being knowledgeable about food has long given Brazilians an aura of cosmopolitanism, but being knowledgeable about the ingredients, techniques, origins, and preparation empowers them in a distinctly new way. Food and cooking have become significant attributes to a middle- and upper-class life.

References

Abala, K. (2007), *Beans: A History*. Oxford/New York: Berg.
Anonymous (2014), "A New Passion for Cooking." *Comer no Brasilia Revista*, August 2014.
Barbosa, L. (2012), "Rice and Beans, Beans and Rice: The Perfect Couple." Pp. 109–119 in R. Wilk and L. Barbosa (eds), *Rice and Beans: A Unique Dish in a Hundred Places*. Oxford/New York: Berg.
Embrapa (2009), *COZINHA Experimental da Embrapa Arroz e Feijão*. Embrapa: Brasilia. https://www.embrapa.br/busca-de-publicacoes/-/publicacao/204775/cozinha-experimental-da-embrapa-arroz-e-feijao. Accessed January 30, 2015.
Fajans, J. (2012), *Brazilian Food: Race, Class and Identity in Brazilian Cuisines*. Oxford/New York: Berg.
Flavors of Brazil http://flavorsofbrazil.blogspot.com/2010/04/utensils-clay-cookware-from-goabeiras.html. Accessed January 15, 2015.

Fry, P. (2001), "Feijoada e *Soul Food 25* anos depois." Pp. 25–54 in N. Esterci, P. Fry, and M. Goldenberg (eds), *Fazendo Antropologia no Brasil*. Rio de Janeiro: DP&A Editora.

Goldstein, D. (2003), *Laughter Out of Place: Race, Class Violence, and Sexuality in a Rio Shantytown*. Berkeley: University of California Press.

Goody, J. (1982), *Cooking, Cuisine and Class: A Study in Comparative Sociology*. Cambridge, UK: Cambridge University Press.

Instituto do Patrimônio Histórico e Artistico Nacional http://portal.iphan.gov.br/portal/montarPaginaInicial.do;jsessionid=BE50CC57FECE9DF31B7BFC49E929BFF4. Accessed January 30, 2015.

Naccarato, P., and Lebesco, K. (2012), *Culinary Capital*. New York/London: Berg.

Scheper-Hughes, N. (1992), *Death Without Weeping: The Violence of Everyday Life in Brazil*. Berkeley: University of California Press.

Veja Comer & Beber Brasilia (2014), "Novo Culinaria, Brasilia." In *Veja Comer & Beber Brasilia*. August.

Wilk, R. and Barbosa, L. (eds) (2012), *Rice and Beans: A Unique Dish in a Hundred Places*. New York/London: Berg.

9

Global dimensions of domestic practices:

Kitchen technologies in Cuba

Anna Cristina Pertierra, University of Western Sydney

Introduction

When people think about the Cuban revolution, the greatest transformations that socialism has brought to the island are usually imagined to have taken place in public and civic spaces. Popular ideas of socialist Cuba invoke such images as the public square full of cheering crowds, while the most famous reforms have occurred in such civic spaces as schools and hospitals. However, the ongoing project of a Cuban socialist revolution—albeit one that has repeatedly adapted and mutated in the face of global political and economic pressures—has also been materialized in crucial ways through the most banal and intimate practices and objects of the domestic realm. Everyday household spaces have often been the location in which Cuban citizens have most keenly felt the impacts of change in the more than fifty years of reform for which their nation has become famous.

While debates about the merits and challenges of the Cuban revolution have most frequently focused on questions of democracy, market reform and international relations, for most Cubans the merits and challenges of their revolution are most keenly felt in terms of how such daily practices as shopping, cleaning, cooking, and caring for family members can be made more or less difficult. In this chapter, we will consider how the space of the kitchen has been shaped and reshaped by Cuban socialism, especially in the context of the post-Soviet era, which obliged the Cuban government from the 1990s onwards to attempt new policies following the collapse of crucial economic and political support from the former socialist bloc. Elsewhere I have examined in some detail how everyday consumption strategies and household practices were radically disrupted in the post-Soviet economic crisis that was commonly referred to in Cuba as the "Special Period" (Pertierra 2011).

In this chapter I will focus more specifically on how the kitchen has continued to play a central role in the politics and economics of Cuba since the beginning of the twenty-first century. In particular, the 2006 national policy known as the "Energy Revolution" is indicative of recognition by Cuban policymakers as well as everyday residents that kitchens are a space in which most intimate practices of home-making can also form important sites for political negotiation and even nation-building. It is through the use of kitchens, and the incorporation of seemingly apolitical technologies such as kitchen appliances, that Cubans come to experience their participation—both positively and negatively—in the networks of consumption and distribution that mark Cuba as distinctly non-capitalist. Despite these notable differences to other Latin American contexts, it is however also important to acknowledge that the identities and practices constituted by Cubans—and especially Cuban women—through their kitchens retain many similarities to the experiences of women in other parts of the Americas, and indeed across the modern world. As can be seen elsewhere in this volume, Cuban women share with many others an experience of using their kitchens as a space through which they overcome considerable adversities to produce households and families that are nourished and sustained nutritionally, as well as in other ways. The cultural significance with which kitchens are imbued is precisely that which makes them a useful site for interventions by the Cuban state.

The politics of kitchens

While kitchens may not seem at first glance to be overtly politicized spaces, scholars from across the social sciences have in recent decades increasingly acknowledged the kitchen not only as a space in which politics is produced at the micro level, but also a locus from which politics can be played out at national and international levels. Feminist analyses in this field have focused particularly on the role that kitchens play in women's lives, and the potential power of kitchens as spaces from which gendered politics can be both produced and mobilized (Schwartz Cowan 1996; Murcott 1989; Johnson 2006). Historians and geographers have considered how kitchens can be bearers of modernity across a number of different settings. While such models of modernity may be embraced or resisted in unpredictable ways, this scholarship acknowledges that across many cultural and economic contexts kitchens operate as spaces in which women's power is central to the community. Here, women are acknowledged as leaders of political and family networks. Kitchens are therefore important potential agents of change, and this has been recognized by institutions and governments as much as by the users of kitchens themselves, who have ensured that the communal kitchen programs such as those found in Bolivia and Peru, for example, are frequent targets of development initiatives and training schemes (Shroeder 2006; see also Johnson 2006).

Perhaps the clearest example of such kitchen politics writ large across the global stage is that of the so-called "kitchen debates" between Richard Nixon and Nikita Khrushchev at the 1959 Moscow World Fair. As historians of technology Oldenziel and Zachmann recount in the opening pages of their edited collection on Cold War kitchens (2009), a tour of the American kitchen exhibit at the World Fair prompted an infamous exchange in which the two statesmen boasted competitively about the technological superiority of their respective countries. In this exchange, the kitchen was invoked not only as a site of competitive material advancement, but as a site in which citizens could achieve competing kinds of modernity—whether Soviet or US—on a much more personal scale than that of the technological innovations associated with the space race. Indeed, Oldenziel and Zachmann's collection demonstrates that in the Cold War era, politicians across Europe "considered kitchen appliances as the building blocks for the social contract between citizens and the state" (2009: 3; see also Fehérváry 2002: 384). By the late 1950s, two distinctly different visions of Cold War modernity—socialist and capitalist—were increasingly being played out with reference to the material culture of domestic life. While the US emphasis was upon individualized consumer items as technological advancement, in the Soviet bloc technological developments emphasised the value of communal technologies such as public transportation, notwithstanding the relative widening of individual consumption options that were introduced in the Khrushchev era. Across socialist Europe, the mid-twentieth century saw a range of government initiatives which acknowledged how important it was to offer citizens the means to improve their domestic spaces in terms of both a modern aesthetic and a contemporary ideology of labor-saving and convenience (Crowley and Reid 2002). Krisztina Fehérváry's excellent study of Hungarian kitchens draws attention to the role that consumer goods and material culture played in the Cold War competition to realize a kind of domestic socialist modernity:

> While state-socialist regimes were known for their inability to produce abundant consumer goods and high standards of living, I have argued that

it was nonetheless under these very regimes that mass consumer society and modern consumer subjectivities emerged – and with it, particular relationships to material culture. Reinforcing rather than contradicting widespread temporal and moralizing discourses of modernization, the socialist state prioritized the material project of becoming modern over other modes of attaining social responsibility.

FEHÉRVÁRY 2002: 394

1959, the year of the Moscow "kitchen debates," was also a landmark year in Cuban history, with the ousting of the Batista regime and the declared triumph of the Cuban revolution. That this revolution occurred at precisely the moment in which consumer modernity was being played out through Cold War competition is telling, because within Cuba, deeply polarized experiences of American modernity were themselves an important element in the foment of political unrest. While middle-class Cubans enjoyed a lifestyle that was deeply integrated into the products and practices of US consumer culture, such experiences were by no means accessible for all of the Cuban population. It was precisely such economic and cultural polarization that the following fifty years of Cuban policy-making, through various stages and adaptations of state socialism, has attempted to eradicate. Therefore, alongside the reform of schools and hospitals, Cuban housing policy was also reformed and the Cuban version of domestic modernity was increasingly seen as a goal and achievement of the revolution. Deprived citizens in both urban and rural communities became increasingly integrated into household practices that required running water, electricity, and cooking gas, along with the appliances and objects that are dependent upon such utilities. In some cases, housing was redistributed or subdivided to move previously marginal Cubans into housing that had formerly belonged to elites, while the 1960s, 70s and 80s also saw the development of major housing developments to relocate those whose housing had previously been poor. Despite such initiatives, there have been considerable housing shortages in city areas over the past fifty years, and internal migration has been carefully controlled, although since 2014 some important reforms in housing legislation have been having an impact on Cubans' opportunities to buy, sell, and move houses. One result of this remarkable stability of population is that many families still live in the same houses that their parents or grandparents previously occupied. Without a formal property market, houses are frequently rebuilt, subdivided, or even swapped to accommodate the changing needs of households, but such changes rely more heavily on the recirculation of long-existing resources and materials than on an accelerated accumulation of new houses and housing materials.

Cuban domesticity has not only been changed in terms of housing policy, but also in terms of the interiors and contents of the average Cuban home. Thinking about what Igor Kopytoff (1986) would describe as the "cultural biographies" of kitchen technologies—along with other technologies such as televisions and electric fans—can tell us a great deal about how Cuban citizens' intimate lives and practices are bound up in the political and economic relations that have changed while the Cuban government has grappled with transitions into and out of Soviet-style socialism. Thus, in Cuba, "kitchen politics" is at times the result of very conscious and specific shifts in economic and consumption policy at the state level. How people in turn respond to such official politics across their diverse experiences, identities, and positions, suggests that those citizens "on the ground" are just as aware as state policy-makers of the crucial role that kitchen technologies play in shaping and maintaining daily routines and attaining or securing standards of comfort and life satisfaction.

Although the starting point of this chapter has considered the politics of Cuban kitchens from a somewhat abstract perspective, I will now turn to material drawn from the much more fundamental experiences and practices of Cuban women who have participated in my intermittent ethnographic research on consumption and everyday life in the city of Santiago de Cuba from 2003–14. As we shall see, throughout the moderate economic reforms that Cuba has undergone over the past decade, some households and families have been able to secure a level of relative affluence and domestic comfort clearly measured by the strong presence of domestic appliances in their everyday lives. But for others whose access to new forms of income in the changing economy has been limited, reliance upon state-run schemes to secure key domestic appliances is as strong as ever. Having discussed two brief examples of how Cuban kitchens are equipped today, I turn to the specific history and outcomes of the nationwide distribution policy launched in 2006, known as the "Energy Revolution," in order to demonstrate the ongoing relationship between state and kitchen that underpins significant features of contemporary Cuban society.

Cuban kitchens

Although the first decades of the socialist revolution greatly diminished the gap between rich and poor in Cuba, Cubans whose parents and grandparents were middle class before the revolution have often been able to leverage the legacy of their family's former material affluence in meaningful ways (Pertierra 2011). In the case of kitchens, members of the middle classes in the capitalist era already possessed modern kitchens with large refrigerators, wide countertops, installed stoves with ovens, and plumbing. For families living in such houses today, the enduring quality of these pre-1960s kitchens is recognized as being a great advantage in running the home. Indeed, before the 2006 "Energy Revolution," it was common for the refrigerators in such houses to have been used since before the revolution (although admittedly repaired many times), as the US-manufactured Westinghouse refrigerators are widely admired for their large capacity and reliable engineering. Households able to draw from a bank of material goods left over from the pre-revolutionary era also continued to acquire new appliances in the Soviet and post-Soviet periods, such that it is not surprising to find multiple refrigerators in the households of better-off families, or to hear of older refrigerators being sold through informal networks once people are able to upgrade to a newer model. Thus, the kind of kitchen installed in pre-revolution dwellings has remained a building block upon which certain Cubans have been able to create a comfortable domestic life today.

The ability to generate a cash income, whether through particular job opportunities or (more frequently) through family remittances, has played an even more crucial role, however. For more than twenty years, Cuba has operated a dual economy. This means that most public-sector workers receive a salary equivalent to an average of $20 a month. It is only those with other means of income—whether from small businesses, cash tips from tourism, selected professional positions, or family remittances from abroad—who are able to access the many consumer goods available in Cuba at prices that outstrip those across the rest of Latin America. Unsurprisingly, such a dual economy has led to an economic repolarization of Cuban households, and this widening gap is often most sharply felt in the domestic realm. Refrigerators, a key item in the contemporary Cuban kitchen, cannot be bought informally for less than $300, while prices in official stores can be considerably higher. Consequently, refrigerators are an item—and one considered essential for the normal functioning of a household—which is almost impossible to afford, whether new or used, by most Cubans. Even a kitchen appliance such as a blender—sold in the stores for between $30 and $50—must be saved for over time unless it can be bought as a gift from a relative with an income. Within this context, categories of kitchen appliances that have never been available through Soviet-era acquisition schemes and which did not exist before the revolution—the microwave being a prime example—suggest by their mere presence in a Cuban kitchen that the household has been a relative "winner" in the post-Soviet era, whether through work opportunities or family migration. Conversely, one can see merely by entering the kitchen of a pensioner or public-sector worker that the lack of key appliances therein signifies a lack of access to the cash economy.

A brief comparison of two women and their material surroundings can show how Cubans who are officially in the same demographic categories experience different kinds of kitchens due to their position within the post-Soviet economy. Maricel and Miriam are both retired women who lived alone in 2014, and who had formerly worked in the cultural and educational sectors their entire lives. Both women are respected and respectable members of their community, living not far from one another in the city of Santiago de Cuba. Maricel, divorced and in her early sixties, has two adult children living abroad, as well as numerous cousins and siblings who have emigrated to the United States over the past fifteen years. Maricel relies on her Cuban state pension for her everyday needs, and does not rely on her children or emigrant relatives to support her financially with regular money transfers. However, when there is a major expense, she has a network of people upon whom she can call for help. Further, the material benefits of her transnational relations can clearly be seen in the contents of her kitchen. Maricel's refrigerator was given to her by her brother before he left Cuba in the early 2000s. She has a microwave, purchased with money given to her by a relative. In addition to her appliances, Maricel's kitchen also features smaller objects brought back to Cuba from Europe, where she has spent time visiting relatives in recent years. Such seemingly humble objects, almost impossible to find locally, as tin openers, garlic crushers, chopping boards, or colanders, sourced in international stores like Ikea, are a tell-tale sign in Maricel's kitchen of the international ties that boost her everyday quality of life at home. While Maricel's kitchen is small, it is well equipped.

Miriam's house, while by no means deprived in the context of Cuba, provides a useful contrast to the relative comfort of Maricel's kitchen. Widowed for more than twenty years, Miriam is in her seventies and worked full-time until very recently. Like Maricel, Miriam also relies on a state pension for her everyday expenses, but has no children and none of her close relatives lives abroad. Miriam's lifelong integration into community and labor-related activities means that she has been able to acquire domestic goods through many of the socialist distribution schemes that dominated in the Soviet era, and which still have a legacy in Cuban homes today. Thus, Miriam has a "home telephone," although this is actually a community telephone used by neighbours in her street. She was also able to purchase or be awarded kitchen items including crockery, cutlery, and tableware during her working life, and as these goods are well cared for Miriam does not have to replace or update them. However, Miriam is not able to buy items that require some form of private income, such as microwaves. The most noteworthy development in recent years to have affected the contents of her kitchen, therefore, was the Energy Revolution, through which her refrigerator was acquired alongside a range of other appliances (Miriam's acquired her previous refrigerator, which was Cuban-made, in the 1970s and it regularly malfunctioned). While all Cubans, including those like Maricel who have other resources beyond the state, have benefited from the Energy Revolution to some degree, it is people like Miriam, whose economic wellbeing depends upon the socialist economy, for whom this policy was designed.

The 2006 "Energy Revolution"

In a characteristically long speech delivered in the Western province of Pinar del Rio in 2006, Fidel Castro emphasized the importance of launching a group of new policies under the banner of the *Revolución Energética* or Energy Revolution. Predicting that future Cubans would mark this as a historical turning point, Castro declared that "There will be a before and an after [when speaking] of the Cuban energy revolution, from which we will be able to learn useful lessons for our population and for the populations of the rest of the world" (Castro 2006, author's translation). While a swathe of energy-saving reforms were announced across industrial sectors and in the energy infrastructure nationally, for most Cuban citizens the Energy Revolution was principally experienced through a nationwide program that provided a range of domestic appliances to every household. Items bought on instalment or traded in for newer models included stovetop pressure cookers, rice cookers, electric pressure cookers (popularly known as the *Reina* or "Queen" for their versatility in the kitchen), electric hotplates, portable water heaters, electric fans, televisions, light bulbs, pots and pans, refrigerators and thermostats, as well as the assorted smaller parts needed to replace or maintain various domestic technologies (Castro 2006). Brigades of social workers and other sectors of the young population were mobilized through every region of Cuba, going from house to house to produce inventories of which appliances people already had, which would be traded in, and which would be bought new at subsidized prices. While participation in the scheme was not compulsory, it was very highly encouraged. For most Cubans getting involved was attractive in any case, as it offered access to key domestic technologies at more affordable prices than were normally available. All households encountered in this research purchased on instalment smaller items like pots and pans, water heaters, and electric hotplates. Larger items, such as refrigerators, were typically traded in, except by those households who had new models already: Maricel, for example, refused to trade in the refrigerator gifted by her brother as she calculated that it was better value than the replacement refrigerator for which she would have had to pay. But most people, like Miriam, decided that although she would have to make payments for her refrigerator for a much longer time than for any of the smaller items she received, the sacrifice was worth it in order to upgrade her kitchen.

The Energy Revolution was not without its problems: shortly after appliances were distributed stories began to circulate about poorly produced items breaking down or never working at all. Further, while the emphasis on electric appliances (replacing gas, kerosene, or even wood-fired technologies) was intended to save power across the nation, it made citizens especially vulnerable when blackouts occurred. For most Cuban families, however, the policy resulted in a welcome—if not dramatic—increase in the convenience of their kitchen routines. Above all, the distribution of the *Reina* electric pressure cooker was met with great enthusiasm by the women who were typically in charge of household meals. Recent research in June 2014 found that, around eight years after the policy was introduced, the *Reina* remained an important appliance and was used in on a near daily basis. The *Reina* is an excellent labor- and time-saving device that can be used to cook almost any of the basic ingredients in a typical Cuban meal (such as the rice, beans, braised meats and root vegetables used in soups and stews). When bought new in the free market stores, such electric pressure cookers can cost more than US$100. In contrast, while the *Reinas* were the most expensive of the cooking appliances in the Energy Revolution scheme, they were distributed for the price of around US$12. As a result, even when some people's Energy Revolution-sourced *Reina* starts to malfunction, efforts are made to replace the parts as the appliance has come to occupy such a central part of everyday cooking practices.

The Energy Revolution initiatives demonstrate that the Cuban government continues to seek methods for distribution that retain the communitarian ideological basis of socialism even amidst an era of growing economic disparity, and also remind us that socialism has long involved a concern with domestic consumption as a locus for emotional wellbeing and, indeed, political participation. However, even within the context of egalitarian policies like the Energy Revolution, tensions and contrast can clearly be seen in terms of how the distribution of domestic appliances has had lasting effects for the population. In a seeming paradox, by 2014 the residents of Santiago de Cuba who were still making best use of their Energy Revolution appliances were generally those who were already among the materially most privileged. The families for whom the 2006 appliances were additional to their existing equipment had managed to keep operating multiple technologies, either rotating them when an appliance had to be repaired, or saving certain appliances as "reserves" for the following years. By contrast, the households that did not have a *Reina*, portable water-heater, or electric stovetop prior to Energy Revolution subjected their appliances to much heavier use. As a result, many of these appliances had broken, and in such households it is today rare to find all of the technologies acquired in 2006 still in use.

Conclusion

In her oft-cited collaboration with Michel de Certeau and Pierre Mayol on the practices of everyday life, Luce Giard observes that "*doing-cooking* is the medium for a basic, humble, and persistent practice that is repeated in time and space, rooted in the fabric of relationships to others and to one's self" (1998: 157). Through the repetition of such practices, and the relationships they enable, Giard argues, women participate in a most connected and intimate way in the shaping and reproduction of families and also of societies. Giard writes about this "Kitchen Women Nation" in modern France as something that is under-acknowledged and subtle in its power (1998: 155). But in the context of Cuba there is nothing subtle or unacknowledged about the importance of everyday kitchen practices, and the obligations of the Cuban state in providing avenues for such everyday kitchen practices to be maintained if not improved. Policies such as the 2006 Energy Revolution demonstrate in a very stark manner the direct relationship that Cubans experience between the structures of their political system, the consequences of their economic precariousness, and the spaces and materials of their domestic environment. It is by no means an exaggeration to suggest that the most important ways in which Cuban citizens measure the ongoing validity of their political system relate to their experiences of, frustrations with, or aspirations toward, consumption and material culture. As Fehérváry posits from the perspective of post-socialist Europe, "Material culture, indeed, is emerging as a particularly revealing site for the investigation of the effects of the fall of state socialism on practices, values and subjectivities" (2002: 370). While in the Cuban context state socialism has not fallen, it has certainly morphed; to understand how these transitions and transformations are taking place in twenty-first century Cuba, there is no better place to look than the kitchen. It is in kitchens and other household spaces that Cubans, most often women, are quite literally experiencing, through their most basic material interactions with pressure cookers, refrigerators, and other items, the sporadic benefits as well as the recurrent challenges of the post-Soviet economic system. Through these seemingly banal interactions with domestic appliances, one can clearly evaluate how economic polarization, a complex dual economy, the shortages of sanctions, and the material legacies of former political eras constitute diverse economic positionings of families and households in the Cuba of today. Viewing politics through the kitchen, therefore, is a valuable alternative to understanding policies "from above" or even "from outside," as it more effectively captures the cultural realities of politics as it is experienced in the everyday.

Acknowledgments

Research for this article has been supported by a number of schemes and fellowships. Fieldwork in 2003–04 was funded by the Royal Anthropological Institute of Great Britain and Ireland, with follow-up fieldwork in 2006 funded by the University of London Central Research Fund. Between 2008–14, research in Cuba on domestic consumption more broadly was undertaken while completing other research at the University of Queensland Centre for Critical and Cultural Studies. This research was funded in large part by an Australian Research Council Federation Fellow project awarded to Professor Graeme Turner (2008–11) and an Australian Research Council Discovery Award and Australian Postdoctoral Fellowship (2011–14). In 2014, supplementary research in Cuba was funded by the University of Queensland Faculty of Humanities and Social Sciences. Research assistance was provided during 2014 by Yoelbys Trimiño Martínez and Ivonne Sanchez Noroña. My sincere thanks go to these institutions and individuals for their support.

References

Castro, F. (2006), "Habrá un antes y un después de la Revolución Energética de Cuba." *Granma* 18 September. Available online: http://www.granma.cu/granmad/secciones/ noal-14/decuba/cuba-revolucion-energetica.html. Accessed January 31, 2015.

Crowley, D. and Reid, S. E. (eds) (2002), *Socialist Spaces: Sites of Everyday Life in the Eastern Bloc*. Oxford: Berg.

de Certeau, M., Giard, L., and Mayol, P. (1998), *The Practice of Everyday Life Volume 2: Living and Cooking*. Trans. T. J. Tomasik, Minneapolis: University of Minnesota Press.

Fehérváry, K. (2002) "American Kitchens, Luxury Bathrooms, and the Search for a 'Normal' Life in Postsocialist Hungary." *Ethnos: Journal of Anthropology* 67(3):369–400.

Johnson, L. (2006), "Hybrid and Global Kitchens—First and Third World Intersections (Part 2)." *Gender, Place and Culture: A Journal of Feminist Geography* 13(6):647–652.

Kopytoff, I. (1986), "The Cultural Biography of Things: Commoditization as Process." Pp. 64–91 in A. Appadurai (ed.), *The Social Life of Things: Commodities in Cultural Perspective*. Cambridge: Cambridge University Press.

Murcott, A. (1989), "Women's Place: Cookbooks' Images of Technique and Technology in the British Kitchen." *Women's Studies International Forum* 6(1):33–39.

Oldenziel, R. and Zachmann, K. (2009), "Kitchens as Technology and Politics: An Introduction." Pp. 1–33 in R. Oldenziel and K. Zachmann (eds), *Cold War Kitchen: Americanization, Technology, and European Users*. Cambridge, MA: The MIT Press.

Pertierra, A. C. (2011), *Cuba: The Struggle for Consumption*. Coconut Creek: Caribbean Studies Press.

Schwartz Cowan, R. (1996), "Technology Is to Science as Female Is to Male: Musings on the History and Character of Our Discipline." *Technology and Culture* 37(3):572–582.

Shroeder, K. (2006), "A Feminist Examination of Community Kitchens in Peru and Bolivia." *Gender, Place and Culture: A Journal of Feminist Geography* 13(6):663–668.

PART THREE

Recreating tradition and newness

10

Recipes for crossing boundaries: Peruvian fusion

Raúl Matta, University of Göttingen

Introduction: Contextualizing Peruvian fusion

Peruvian cuisine is arousing great enthusiasm across the country today. Never before have Peruvians talked so much about food. They take pride in "their" cuisine, think about its international potential, and consider successful local chefs to be artists, if not superstars. The omnipresence of food in the public sphere is the result of what is known as the "Peruvian gastronomic revolution:" a phenomenon which originates from the early 1990s, was developed further in the mid-2000s, and has shaped the conditions developing the country's fine-dining market (Lauer and Lauer 2006). This phenomenon is strongly related to recent changes in the country's society and politics. Not long ago, Peru was not a particularly attractive destination. Civil war and a significant economic crisis were the backdrop to the 1980s, but now that the war has ended, both tourism and the economy are strong. The application of neoliberal policies during the 1990s increased the purchasing power in the cities, mainly in the capital, and placed Peru as a nascent actor on the global scale. The growing economy and the arrival of global cultural patterns prompted processes of class differentiation through new tastes and consumption trends (Matta 2009). In such a context, the interest in gastronomy, one expanding aspect of urban economies, is anything but surprising.

These contextual factors matter as much as the changing trends in gastronomic business models and activity worldwide. The move of Peruvian food from the domestic into the public sphere was facilitated by the professionalization of cooking in Lima, both through the creation of occupational associations and the increasing attractiveness of gastronomy as a profession. The latter fact is linked to the rise in social position gained by chefs during the 1970s and 1980s in Europe and the United States, when the key players behind French nouvelle cuisine and California's fusion cuisine launched their own restaurants (Drouard 2004; Gordon 1986). Since then, the gastronomic business has grown in importance as a key component of the cultural industry, while cookery has increasingly gained respect. Although gastronomy has, over the years, become a very competitive market in which compensation and reputation is concentrated in a few top performers, the self-marketing strategies pursued by chefs (and which made media stars of some) have encouraged both in Peru and beyond the opening of restaurants and an increase in publications, TV shows, and culinary schools. Gone is the notion of cookery as a difficult career associated with subordinate, servile tasks—although, in practice, this was never really the case. On the contrary, qualified culinary skills are now recognized as accomplishing important social and economic functions.

Peruvian cuisine is seen as a legitimate cultural field and as a profitable economic activity. Furthermore, the cooking profession's increasing reputation has challenged the sense of social prestige of the very conservative Peruvian upper class, which finally understood that in current times prestige could also be reached by becoming a cultural and/or media personality. Indeed, the leaders of the Peruvian gastronomic revolution are typically members of well-off families of European descent who have had the chance to receive international culinary training during the 1990s and early 2000s (Matta 2010). The social background of these cooks plays a crucial role in their attempts to become elite chefs, since it allows them to speak with authority about their reinterpretations of Peruvian cuisine. Indeed, they have not had to face social or cultural barriers when they offer novel creations, unusual ingredients, and stylistically daring performances to their clients and social peers: their heritage has made the task easier and made people less suspicious of them. Therefore, it would not be unfair to say that their first successes are as much due to their privileged social positions as their individual skills.

As the culinary industry grows globally, so the promotion of a variety of cuisines serves as an important way of expanding the potential audience. Current developments in the gastronomic industry and press entice foodies from around the globe to try out new restaurants and search for new cuisines and tastes. New markets for food products, cooking technologies, and culinary skills and discourses emerge. Peruvian chefs and food entrepreneurs have moved rapidly into available "ethnic" and "exotic" niches characterized by the use of local or "indigenous" food from the interior of the country. In fact, Peru is a country whose population of European migrant origin views its indigenous, Amerindian population as "ethnic."

Recent narratives of *terroir*, heritage, cultural identity, and authenticity at play in the food and tourism industries (Abarca 2004; Poulain 2012) have also stimulated the latest transformations in Peruvian restaurant cooking. New trends such as Nuevo Latino (Fonseca 2005) and Nordic Food (Tholstrup Hermansen 2012)—as well as the rise of local food activism, such as the Slow Food movement—have paved the way for highly trained cooks to draw on "native" ingredients and traditional food as a means to foster the versatility of Peruvian cuisine. Key in this endeavor is the use of Andean and Amazonian elements. As explained elsewhere (Matta 2013), Peruvian chefs constantly reappropriate and resignify food long seen as backward and old-fashioned by social elites by using haute cuisine techniques and aesthetics. Therefore, items kept away from elite tables, such as guinea pig, *paiche (Arapaima gigas Cuvier,* a large fish from the Amazon river*), arracacha (Arracacia xanthorrhiza Bancrof,* a root vegetable similar to carrot and celery), or *cushuro (Nostoc sphaericum Vaucher* a blue-green spherical algae) are obtaining a higher status. This sort of "food gentrification" is achieved across a number of steps: first by removing an item from any prior context—thus neutralizing its indigenous and lower-class characteristics—by then identifying some of its desirable attributes; and finally by connecting it with elements from other culinary cultures, in particular those which already enjoy global recognition. This is indeed how Peruvian cuisine has become more widely known and nourished national and international expectations (Fan 2013; Fraser 2006; Matta 2011; Mclaughlin 2011).

It is important to note here that although media and public opinion refer to recent chefs' creations as "Peruvian cuisine," closer examinations of restaurant menus and websites reveal that these examples barely fit into this category. Indeed, these new interpretations do not appear in Peruvian cookbooks and food writings, and they are not served in humble restaurants or in homes. As we will see later, although chefs make more use of the food diversity existing in Peru, their major goal is to conform to international standards and not necessarily to reassert Peruvian cuisine (Matta 2010). For this reason, I refrain from using the term "Peruvian cuisine" to refer to this culinary trend. Instead, I prefer to speak of a fusion model, or namely, of "Peruvian fusion," which I define broadly as the combination of more or less recognizable "Peruvian" culinary forms (recipes, ingredients, and techniques) and other forms coming from outside the borders of the country. The essence of Peruvian fusion could be summarized as fusion cuisine with "Peruvian DNA."

This chapter explains the background to Peruvian fusion. It shows how cooking techniques, technologies, and knowledge are "elevating" indigenous and formerly disregarded food ingredients from basic cookery to gastronomy and from local to global. It also aims to demonstrate how, through the interweaving of chefs' practices and discourses, Peruvian fusion contributes to challenging representations of "the inedible" that have been rooted in long-standing racial and social prejudices. In fact, Peruvian fusion has to be understood as an outcome of the global forces of food and tourism industries that, as they expand relentlessly, allow—if not directly promote—processes that neutralize "negative exoticisms" and instead propose "positive exoticisms."

The setting for Peruvian fusion

This chapter is based on observations and interviews conducted in August 2011 at Central and Mayta, two fine-dining restaurants in the vibrant district of Miraflores in Lima. These venues are among the most respected in the Peruvian fusion movement and are perfect examples of the cosmopolitan ambitions of this trend. Entering these places is similar to entering fashionable and upscale restaurants in major cities such as New York, London, Paris, or Hong Kong. The influence of minimalist architecture and design is noticeable; a simple and polished "look" dominates.

I contacted chefs Virgilio Martínez (Central) and Jaime Pesaque (Mayta) by email and asked them to cook one representative dish from their work for the purposes of research at the intersection of food, cultural heritage, and identities.[1] The idea was to film and take pictures of their performance, as I wanted to obtain better technical insight into the valorization of indigenous food within the framework of Peruvian fusion. Of course, the cooks were at liberty to choose which elements to use.

I was surprised and pleased when both chefs kindly accepted my request. I obtained their approval to film and photograph in the morning, before the service began and after the *mise en place* (i.e. the kitchen's preparatory work) was done. That means that the focus was on the assembling of the ingredients on the plate and on the cooking, which is usually done to order. After recording the preparations, nonstructured interviews followed so that I could find out more about the origins of the dishes, the skills involved in their preparation, and the ways the chefs responded to customers' demands. These interviews proved to be significant as these cooks were trained in schools in which gastronomic background codified in France is taught and consequently used to articulate ideas about taste and to classify dishes.

The researcher's background: Eating the inedible

A few days before our meetings, the chefs told me which dishes they planned to prepare. Martínez would use a selection of Andean tubers, and Pesaque guinea pig. Their choice of ingredients coincidentally connected with one of my most cherished previous food experiences, although this had occurred in a completely different context. In 2007, Berbelina, my family's housemaid, invited me to eat *cuy* (the name by which guinea pig is known in Peru) with her and members of her family, migrants from the northern Andes who arrived in Lima in the early 1980s. Berbelina was the person who helped my parents bring me up, and naturally I was thrilled to join her. I met her in the Lurín district on the outskirts of Lima, at the home of a woman who has gained a good reputation for cooking guinea pig. This woman rents out her dining room and culinary skills for events in exchange for thirty *soles* (around US$10) per guinea pig, for a minimum of six animals. Although the price included some bottled beer, it was nonetheless expensive for people on a low income, such as Berbelina's family. Nevertheless, they agreed to pay, as it was a special occasion in which only healthy, full-grown guinea pigs would be served. From our group of seven, only one of us knew exactly how to get to the venue, and so we agreed to meet in two stages. First we met for breakfast at 10.00 a.m. at Berbelina sister's place, a small one-room studio in La Victoria, a modest and working-class district of Lima. We would then all leave together for Lurín. The breakfast was composed of Inca Kola, an extremely sweet and very popular soft drink in Peru, and two sweet-potato sandwiches (thick slices of slightly fried sweet potato inside bun-like bread) per person. This first meal somehow confirmed to me a long-held assumption about sweet potatoes which, I learned afterwards, was widely shared in other developing countries: sweet potato is commonly eaten by poorer people. The crop's "image problem" has been reported by researchers in Africa and Asia, and interpreted as a significant constraint to increased sweet-potato consumption (Tsou and Villareal 1982; Woolfe 1992; Low 2011). In Peru, although locals eat sweet potatoes as accompaniments in some dishes (such as *ceviche*) or as snacks (such as sweet-potato chips), the reputation of the crop is still low. For instance, I clearly remember how my own middle-class relatives purchased sweet potatoes only to feed their dogs.

After a one-hour trip in a *colectivo* (minibus) and *moto-taxi* we arrived at our final destination. The dining room was very humble but spacious; a dusty court separated the house into public and private spaces. The kitchen was located around an open fireplace in the court. The guinea pigs had already been prepared by the time we arrived, and the fur, stomach, and intestines removed. All the other organs (liver, heart, kidneys, and lungs) remained in place. The guinea pigs were seasoned throughout with a dressing made of salt, pepper, ground cumin, and ground chilli. Before frying, vegetable oil was heated in a large pan at a high temperature. The animals were then placed in the pan organ-side down. A few minutes later, they were flipped in order to cook the other side. To test how well they were done, the cook tapped the meat applying light pressure with the back of a fork. Once the cooking was complete, she removed them from the oil and placed them in another pan to drain. Finally, the guinea pigs were presented on the plates with extremities splayed, and with claws and head still in place. Boiled potatoes and *ají amarillo* (yellow chili pepper, *Capsicum baccatum L var. pendulum Willdenow*) sauce were served as accompaniments. The etiquette around eating guinea pigs favors the use of hands over cutlery, and indeed at that time, I would never have considered doing anything other. Berbelina's invitation anchored my representation of sweet potato and guinea pig in a context defined by specific socio-economic, cultural, and geographical backgrounds. A few years later, I realized I was mistaken.

At Virgilio Martínez's Central: Making modest tubers "global"

Since their appearance in Europe, potatoes have been the subject of controversy and their history as an edible plant has been turbulent. Indeed Europeans' first contacts with the root vegetable were not successful. For instance, when the Spanish conquistadors first encountered the potato in Peru, they looked down on it as food for slaves (Salaman 1985). Later, in his 1765 *Encyclopédie*, Denis Diderot stated that the potato could never fit into the category of "enjoyable foods." For many years the potato continued to be regarded as food for peasants, poor people, and farm animals (Berzok 2003). However, its versatility and prolific character has made it central to the European diet from the end of the sixteenth century onwards (Lang 2001). As its culinary values were progressively appreciated, the potato crossed all geographical and socio-economic boundaries to finally become one of the world's most significant foods; the United Nations declared 2008 as the International Year of the Potato. However, other modest tubers have failed to repeat the successful story of the potato and remain in the shadows. In fact, in Lima's upper- and middle-class neighborhoods, Andean and Amazonian tubers such as manioc, sweet potato, *oca* (*Oxalis tuberosa Molina*), *olluco* (*Ullucus tuberosus Caldas*), *arracacha*, and many sorts of "native potatoes" still suffer from the prejudices against everything that originates in the countryside.

Virgilio Martínez, still only in his late thirties, is considered the most innovative chef on Peru's gastronomic scene. His work is helping to change disdainful attitudes towards indigenous food. The son of a lawyer, he studied at the Le Cordon Bleu schools in Ottawa and London, and worked in prestigious restaurants in the United States (Lutèce), Spain (Can Fabes), and Colombia (Astrid and Gastón). Back in Lima, he opened his flagship restaurant, Central, in 2008 (as of 2012 he also owns LIMA, in London). In 2015, Central attained outstanding positions in San Pellegrino's restaurant listings: it was ranked first in "Latin America's 50 Best Restaurants" and fourth in "The World's Best 50 Restaurants." Central's website offers some clues about Martínez's gastronomic philosophy:

> Virgilio chooses to approach the diversity of our ingredients in a manner similar to that used by the peoples of the Andes in pre-Hispanic times: through vertical ecological monitoring [. . .] land is perceived not as a horizontal plane but rather vertically, so that it takes advantage of all that the flora and fauna are able to deliver according to the particularities of each ecological system.[2]

Put simply, Martínez positions himself as an adventurer and (re)discoverer of native ingredients as much as a creative chef. Local and international gourmets acclaim his uninhibited approach that combines the use of fresh and unconventional Andean produce with clean, minimalist, contemporary design. However, when we met in 2011, his attitude was a little more cautious.

At that time, the chef admitted that the work of Peruvian fusion forerunners, such as Gastón Acurio and Pedro Schiaffino (Matta 2010), prevented him from challenging a conservative, timid, and unadventurous market. Instead, he affirmed that recently Peru was developing a great pride in its local ingredients. However, he recognized that an important part of his job lay in continuing to tackle preconceived ideas about Andean food and social-class-related food neophobia (Wilk 2009):

> I was serving "Veal Sweetbreads with Pistachio Puree." Then I changed it to "Veal Sweetbreads with Arracacha Puree." But the sales of the sweetbread dish fell drastically. And that's very funny, because if you taste the pistachio puree, you notice it's not as good as the *arracacha* puree; they are two different things . . . We decided to use a local ingredient, but people really struggle to learn. "What is *arracacha*?," they say. But it's tasty, easy to find, and you can also cook it in very sophisticated ways.
>
> VIRGILIO MARTÍNEZ, personal interview, August 18, 2011

The dish Martínez prepared that day, "Tuna *Tataki* and Tasting of Andean Tubers," is another attempt to bring local and "unfamiliar" ingredients to high cuisine and global standards: "What I wanted to present here, is how Peruvian tubers, which are increasingly being used in restaurants, can be related to a global dish." The chef did so by introducing the tubers into the logic of Peruvian fusion; that is, by connecting ingredients from mainstream gastronomy to others whose less recognizable origins are seen as a hindrance for their integration into global circuits of culinary culture.

As in many fusion cuisine dishes, it is possible to highlight the link between the ingredients they use and their assumed places of origin: the tubers, including *huayro* potato (*Solanum x chaucha*), manioc, and sweet potato, the *ají amarillo*, and the Andean mushrooms from Marayhuaca could easily be associated with Peru, while the technique of tuna *tataki*[3] is associated with Japan. Another key ingredient in this dish is Serrano ham, whose Spanish origin has not been questioned. During the *mise en place*, the tubers were cut in three-centimeter cubes and cooked in different ways. The *Huayro* potato was *confit* in olive oil. The manioc was slightly fried in "charcoal oil"[4] to deliver a "grill scent" and smoky taste, and then dipped in squid ink to obtain a charcoal-like color. The sweet potato was first boiled and then *confit* in olive oil. A little hole was dug in every piece of tuber to hold emulsions of *ají amarillo* and Serrano ham. Andean mushrooms and more Serrano ham, both dehydrated and grounded, completed the ingredients in this preparation.

The chef assembled the dish in less than five minutes. He started by drawing a large line of Serrano ham-aioli with the "spoon swoosh" technique. The line extended closely and parallel to one of the longest sides of the flat rectangular plate, made out of stone. He then cut a piece of tuna *tataki* into thick slices, which he placed one by one on the aioli base, forming something similar to a fallen row of dominoes. Next, the *tataki* was dusted with Serrano ham and Marayhuaca mushroom powders. Parallel to the *tataki*, Martínez drew a thick line of *ají amarillo* emulsion, placed the tubers on it, and proceeded to fill them with emulsions: the *Huayro* potato and the "manioc charcoal" were filled with Serrano ham aioli, while the sweet potato was filled with *ají amarillo* emulsion. He used mustard leaves and daikon radish sprouts to decorate the manioc and the *tataki* respectively.

The result was representative of Martínez's early work in Central: innovative but also cautious. Indeed, Central has earned its international reputation by building on a balance between the familiar (the global) and the unfamiliar (the very local). In that respect, it is important to observe the weight accorded to the ingredients in the preparation. The tuna *tataki* is the protagonist: it is the most identifiable item and the source of protein in the dish. The Andean tubers are first and foremost an invitation to expand customers' knowledge of flavors: "The Japanese would be more than happy with only this mouthful of sweet potato; with that [he points out the *tataki*], they're going to get bored because they can find it everywhere," the chef explained. Finally, the Serrano ham triggers a connection between the flavors of the *tataki* and those of the tubers, while acting as a symbolic bridge between the geographical origins of these ingredients. In the end, it is noticeable that the "non-Peruvian" ingredients dominate the "typically Peruvian" ones.

At Jaime Pesaque's Mayta: Guinea pig is rendered "edible"

For at least 500 years, guinea pigs have been used in the South American Andes as food, diagnostic medical devices, divinatory agents, and as part of religious offerings. The role they play in Andean life is so significant that, in some regions, families who do not raise guinea pigs in their households are accused of being lazy or considered very poor (Morales 1994). Such a centrality cannot, however, be explained without taking into account the historical contribution of these rodents to the indigenous diet. Ethnological and archeological research has stressed connections between the nutritional properties of guinea pigs and the food rituality in which they are involved: guinea pigs served in communal *fiestas* in Andean Peru have been historically used to balance the diet when proteins are in short supply (Rosenfeld 2008; Bolton 1979). Therefore, people in the Andes "do not eat *cuys* because there are *fiestas*, but rather they have major *fiestas* when they do because there is *cuy* meat to eat" (Bolton 1979: 246). However, the high regard for guinea pig as food in the Andes has been inversely proportional to its acceptance in urban areas. The work of DeFrance (2006) backs this up: it shows that in the Andean city of Moquegua, in southern Peru, the guinea pig has evolved from familial food to a prestigious restaurant dish while just ninety kilometers away in the coastal city of Ilo, the guinea pig is regarded as a non-edible meat associated with indigenous culture. In Lima's middle- and upper-class areas, the situation is analogous: guinea pig was regarded as food for peasants and *cholos* (a derogatory racial term for people of Indian descent). For Lima's elites as well as for (non-Andean) people from abroad, guinea pig could not be a part of a meal: while the former regarded eating *cuy* as a betrayal against modernity, the latter have only imagined it as a pet, or as a laboratory test subject.

Yet, recent international press coverage point at a change in attitudes. For instance, the United Arab Emirates-based web-journal *The National* listed Gastón Acurio's "Peking Guinea Pig" as one of the "10 Dishes to Try Before To Die";[5] the Globalpost.com enquires about the presence of guinea pig in fine-dining restaurants and the increasing of exports of frozen *cuy* to North America (Tegel 2014); finally, National Public Radio reported on the increasing consumption of the rodent in the United States (Bland 2013). While the long-term presence of Andean migrants is invoked as the primary reason for guinea pig consumption in the United States, its presence in fancy restaurants around the world is explained as the result of Peruvian chefs' reinterpretations of traditional recipes: "Traditionally the 'cuy' was deep-fried or served in a stew on the Peruvian hillsides. But, fortunately for pet lovers, Acurio [. . .]

has disguised it here as Peking duck, serving it sliced up into chunks with a rocoto pepper hoisin sauce. Once it's wrapped up in a purple corn pancake, you won't even think about Fluffy" (see note 5). The previous quote helps us understand how guinea pig is made "edible": basically, it consists of hiding the fact that it is actually a rodent. As discussed above, in the Andes the animal is cooked in its entirety, never dismembered, and served in one piece. Quite often, tourists ask for it only so that they can take photographs. In upscale restaurants, on the other hand, guinea-pig dishes undergo changes in presentation.

Alongside with Gastón Acurio, Jaime Pesaque is one of the Peruvian chefs who are turning *cuy* into a high-end culinary delicacy. His familial, educational, and professional backgrounds are similar to those of the other chefs leading the gastronomic boom in Peru: Pesaque is the son of a well-to-do family, was trained in prestigious culinary schools, and followed an international trajectory. Back in Peru he opened Mayta, his Lima restaurant and headquarters, in 2010. "Mayta" is the Aymara word used to designate a kind, hearty person. Pesaque says that his family's housemaid inspired him to name the restaurant this, as she often spoke Aymara when cracking jokes with him and his brothers. Pesaque, in his mid-thirties, also runs three restaurants in Miami, one in Hong Kong, one in Punta del Este, and one in Oslo. The chef described his cuisine to me as combining "Peruvian soul" and "Peruvian reminiscences" with the "technique and technology to elevate it to the highest level." His philosophy is similar to that of Martínez: to rediscover and revalorize ingredients from Peru's hinterlands. However, unlike Martínez's adventurer-genius guise, Pesaque's approach is not individualistic. He formulates his commitment as a mission to be shared by the new generation of cooks:

> What we, as young people, have to do now and see as our goal is to use these unknown, forgotten, or little-used ingredients and techniques and introduce them into high-end cooking, as for example the *arracacha*. People might never have heard of *arracacha* nor tried it, or maybe they heard the name once and talk about it as if they know what it is. It's a tuber that tastes a little like manioc, a little like potato, but it's sweet. And if you prepare puree with it, the taste is extraordinary. Now I'm going to serve *cabrito* (goat) and tuna with *arracacha*. The tuna goes with roasted Andean tubers. I will use *oca*, *arracacha,* and *olluco*, a small type of Peruvian potato. The goat will be *confit*, put into oil and smoked salt, and then cooked in the oven until it turns brown on the outside. It's all about putting it in a different context. I don't know whether these dishes will be successful, but my mission is to use these unknown ingredients and to put them into gourmet cookery, as is already happening in other countries. For example, like Alex

Atala does it in his restaurant D.O.M. in Brazil, using ingredients from the Amazon region that nobody knew before. But that is his duty as a cook; that is what he has to do.

JAIME PESAQUE,
personal interview, August 8, 2011

Like Martínez, Pesaque uses the example of *arracacha* to illustrate the new use given to local and underestimated food items. Such a coincidence affirms not only the previous resistance to Andean food among urban foodies, but also illustrates the premise of Peruvian fusion. However, in our meeting in Mayta's kitchen, Pesaque put the tubers aside to perform what he calls an "overturning of tradition" (*un vuelco a la tradición*): he prepared "Crunchy Guinea Pig with Creamy Chickpea Tamales" (*Cuy crocante y tamalitos cremosos de garbanzos*) a dish which, he said, was well received.

The new status of *cuy* is accomplished by applying *sous-vide* cooking, a method used in fine-dining restaurants for decades. Food is vacuum-sealed, cooked, chilled, and reheated. Pesaque confided to me that he invested US$14,000 in professional *sous-vide* equipment to cook his meats at a uniform, controlled temperature. The equipment consists of a vacuum oven and an immersion circulator. The prepared guinea pig can be dried in the vacuum oven at a very low temperature for eight hours, or cooked in an immersion circulator for twelve hours at 140° F. After the water has been removed, one finds that the heat has not affected the proteins, ferments, and aromas: vacuum cooking manages to mimic the true moisture content of foods. The cooked piece of meat can then be kept chilled in the refrigerator for as long as several days. During the *mise en place*, the packaged meat is warmed in a water bath for roughly twenty minutes using the immersion circulator. *Sous-vide* cooking allows Pesaque to obtain guinea pig meat meltingly tender while retaining all of its juices; the day I met him, the chef removed every single bone from the meat simply by using his thumb and index finger. He then lightly fried each portion in olive oil, ensuring a texture crunchy enough to invoke the traditional preparation of *cuy frito* (fried guinea pig) but "improving" on it—the animal can be stringy rather than crunchy in some cases. Once plated, the guinea pig was covered by its own juices and accompanied by two small chickpea *tamales* and *salsa criolla* (red onion juliennes, sliced hot pepper, and lime juice) seasoned with tamarind vinaigrette dressing. Pesaque explained that replacing corn with chickpea in the preparation of *tamales*, and lime juice with tamarind in the dressing, gives his dish a distinctive touch. Unlike the *cuy frito,* the "Crispy Guinea Pig" is eaten with cutlery.

Conclusion: Imagining a global cuisine

As seen in the examples above, a new generation of Peruvian chefs uses top-end culinary techniques, technologies, and discourses to "elevate" "native" ingredients from simple cooking to gastronomy and from local to global. The result is Peruvian fusion, a trend that occupies a niche in the increasingly cosmopolitan and "ethnic-friendly" food market. Peruvian fusion signals a more open attitude toward foods that have hitherto been excluded from fine-dining contexts: Andean and Amazonian ingredients are now sources of inspiration for highly skilled chefs and, consequently, new sources of enjoyment for foodies. It has also contributed to a modest improvement in the way the urban elite looks at the rural world (which at least now accepts that it "exists"). Peruvian fusion does not, however, attempt to subvert the well-established (European) gastronomic hierarchies, nor secure parity between mainstream and non-mainstream culinary knowledge. On the contrary, Andean and Amazonian features are mostly subordinated to chefs' personal skills and goals as well as to Western aesthetic canons. Peruvian fusion needs to be understood within the context of global tourism expansion: its development depends largely on the perceptions of local and international gourmets. The cultural and social meanings embedded in ingredients are therefore strategically used to convince, to persuade, to negotiate, and of course, to impress. Chefs know that the frontiers between the acceptable and the non-acceptable may be thin, and thus tread a tightrope, sometimes concealing the uncanny and at other times "embellishing" reality. Cooking in a vacuum and deboning a guinea pig, dying manioc in black, and naming *cushuro* seaweed "Andean caviar" (as Pesaque admitted he did) are tactics that can take place only in the context of cultural and economic competition on the global stage.

Notes

1 Research funded between 2010 and 2012 by the German Federal Ministry of Education and Research (BMDF) in the framework of the research network desiguALdades.net (Freie Universität Berlin / Ibero-Amerikanisches Institut).

2 See, http://centralrestaurante.com.pe/en/ [accessed January 25, 2015].

3 *Tataki* is a typical Japanese preparation in which a piece of marinated meat, fish, or beef, is seared on the outside while left very rare inside, and then sliced.

4 Although there is not reliable information about the origins of "charcoal oil"—which is more precisely oil flavored with charcoal—it seems to have been popularized by Catalonian chef Ferran Adrià in his restaurant El Bulli.

Charcoal oil is obtained by burning the charcoal until it is red-hot, then plunging it into a metal container containing room-temperature vegetable oil. The charcoal flavor infuses the oil.

5 http://www.thenational.ae/lifestyle/food/in-pictures-10-dishes-to-try-before-you-die#1. Accessed January 25, 2015.

References

Abarca, M. E. (2004), "Authentic or Not It's Original." *Food and Foodways* 12:1–25.

Berzok, L. M. (2003), "Potato." Pp. 198–216 in *Encyclopedia of Food and Culture Vol. III*, New York: Scribner.

Bland, A., "From Pets to Plates: Why More People Are Eating Guinea Pigs." National Public Radio, April 2, 2013, http://www.npr.org/blogs/thesalt/2013/03/12/ 174105739/from-pets-to-plates-why-more-people-are-eating-guinea-pigs. Accessed January 24, 2015.

Bolton, R. (1979), "Guinea Pigs, Protein, and Ritual.' *Ethnology* 18:229–252.

DeFrance, S. (2006), "The Sixth Toe: The Modern Culinary Role of the Guinea Pig in Southern Peru." *Food and Foodways* 14:3–34.

Drouard, A. (2004), *Histoire des cuisiniers en France XIXe–XXe siècle*. Paris: CNRS Editions.

Fan, J. (2013), "Can Ideas About Food Inspire Real Social Change? The Case of Peruvian Gastronomy." *Gastronomica* 13(2):29–40.

Fonseca, V. (2005), "Nuevo Latino: Rebranding Latin American Cuisine." *Consumption Markets & Culture* 8(2):95–130.

Fraser, L., "Next Stop Lima." *Gourmet*, August 2006. http://www.gourmet.com/magazine/2000s/2006/08/nextstoplima. Accessed January 23, 2015.

Gordon, B. (1986), "Shifting Tastes and Terms: The Rise of California Cuisine." *Revue française d'études américaines* 27/28:109–126.

Lang, J. (2001), *Notes of a Potato Watcher*. College Station: Texas A&M University Press.

Lauer, M. and Lauer, V. (2006), *La revolución gastronómica peruana*. Lima: USMP.

Low, J., "Sweetpotato for Profit and Health Initiative." Sweetpotato Knowledge Portal, March 10, 2011. http://sweetpotatoknowledge.org/sweetpotato-introduction/ overview/sweetpotato-for-profit-and-health-initiative. Accessed January 24, 2015.

Matta, R. (2009), "Enjeux sociaux d'une consommation 'haut de gamme.' Étude sur les logiques marchandes et sociales au cœur de deux expressions culturelles dans la ville de Lima : l'expérience gastronomique et les fêtes de musique électronique." PhD Thesis, Sorbonne Nouvelle University.

Matta, R. (2010), "L'indien à table dans les grands restaurants de Lima. Cuisiniers d'élite et naissance d'une 'cuisine fusion' à base péruvienne.' *Anthropology of Food* 7, http://aof.revues.org/6592. Accessed January 25, 2015.

Matta, R. (2011), "Posibilidades y límites del desarrollo en el patrimonio inmaterial: El caso de la cocina peruana.' *Apuntes. Revista de estudios sobre patrimonio cultural* 24(2):196–207.

Matta, R. (2013), "Valuing Native Eating: The Modern Roots of Peruvian Food Heritage." *Anthropology of Food* S8, http://aof.revues.org/7361. Accessed January 25, 2015.

Mclaughlin, K., "The Next Big Thing: Peruvian Food." *The Wall Street Journal*, September 10, 2011. http://www.wsj.com/articles/ SB10001424053111904199 404576540970634332968. Accessed January 25, 2014.

Morales, E. (1994), "The Guinea Pig in the Andean Economy: From Household Animals to Market Economy." *Latin American Research Review* 29:129–142.

Poulain, J-P. (2012), "The Sociology of Gastronomic Decolonization." Pp. 218–232 in S. Nair-Venugopal (ed.), *The Gaze of the West and Framings of the East*. Basingstoke, Hampshire: Palgrave Macmillan.

Rosenfeld, S. A. (2008), "Delicious Guinea Pigs: Seasonality Studies and the Use of Fat in the pre-Columbian Andean diet." *Quaternary International* 180:127–134.

Salaman, R. (1985), *The History and Social Influence of the Potato*. Cambridge: Cambridge University Press.

Tegel, S. "Red wine or white, sir, with your guinea pig?" GlobalPost.com, April 25, 2014. http://www.globalpost.com/dispatch/news/regions/americas/ peru/140425/peruvian-cuisine-cuy-guinea-pig. Accessed January 25, 2015.

Tholstrup Hermansen, M. E. (2012), "Creating *Terroir*." *Anthropology of Food* S7, http://aof.revues.org/7249. Accessed January 25, 2015.

Tsou, S.C.S. and Villareal, R. L. (1982), "Resistance to Eating Sweet Potato." Pp. 37–44 in R. L. Villareal and T. D. Griggs (eds), *Sweet Potato: Proceedings of the First International Symposium*, Taiwan: Asia Vegetable Research and Development Center.

Wilk, R. (2009), "Difference on the Menu: Neophilia, Neophobia, and Globalization." Pp. 185–196 in D. Inglis and D. Gimlin (eds), *The Globalization of Food*. Oxford/New York: Berg Publishers.

Woolfe, J. (1992), *Sweet Potato: An Untapped Food Resource*. Cambridge: Cambridge University Press.

11

Forms of Colombian cuisine:

Interpretation of traditional culinary knowledge in three cultural settings

*Juliana Duque Mahecha,
Cornell University*

"*Para ser universal hay que ser local.*"
LA URBANA COCINA, BOGOTÁ, COLOMBIA

What is at stake when interpreting traditional Colombian cuisine? What happens when a *sancocho* soup is deconstructed[1] or *fríjoles* dish—one of Colombia's most popular dishes—is served in "tower-style" (i.e., piled up)? Do they threaten or do they substitute traditional plates or can both variations actually complement each other and exist side by side? If such "reimagined" dishes cost three times as much as the conventional dish, or furthermore if a traditional *fríjoles* dish today costs three times as much as it used to a few years ago (in some establishments at least), how can we explain these changes and why do they turn the relevant items into potentially noninclusive or a more restrictive alternative? In short, does the "modern" version of a prominent local dish challenge traditional cuisine or does it honor the original version?

Change is a sign of movement, of learning, and of the production of knowledge. However, the continuation of certain practices and habits is a must for the survival of particular communities. This constant tension between change and permanence lies at the heart of human endeavor and of what we call culture. If we associate the concept "traditional" with an origin and that which never changes, and in turn the concept "modern" with all things novel and dynamic, then culinary traditions,[2] among others, could be construed as endangered and therefore need to be protected and taught in order to keep them alive. Determining how to do so is where complications and debate arise.

Over the past thirty years we have witnessed a significant global spread of dietary, culinary, and gastronomic movements. This expansion is reflected in the growing interest for such matters on the part of government institutions, private organizations, and citizen groups. It is mirrored in the ever larger number of restaurants, catering and culinary schools, specialty shops, farmers' markets, TV programs, specialized books and magazines, not to mention related festivals and events. In the midst of these processes, the culinary realm in Colombia is changing as well. Thus, efforts to recover, discover (or rediscover), to protect and disseminate, to recreate, to develop, or modernize some dishes are activities currently in play. The growth of the culinary industry in Colombia has been prominently marked by the opening of several fine-dining establishments and other type of restaurants in the last decade.[3] Many of these advertise themselves offering a "native cuisine," a Colombian cuisine based on "fusion cooking," a "gourmet Colombian cuisine," or simply a "new" Colombian cuisine.[4] Other eateries claim to offer authentic traditional cuisine, fresh market fare, or a particular regional cuisine; others still offer home-style cooking or "popular" dishes.

This chapter looks comparatively at the ways in which in such a context, Colombian traditional cuisines are interpreted and enacted in three different urban food settings: fine-dining restaurants; comfort-food restaurants, and food stands at markets.[5] Dining out is becoming more prevalent in Colombia, and combined with the formulation of a "new" national cuisine, seems to be contributing to the consolidation of new senses of belonging. These are mainly defined by the interplay between contemporary global and local processes. The emergent gastronomy is based on native cuisines and culinary traditions, but also comprises cosmopolitan values that frame it in political and aesthetic discourses of modern consumption.

The three culinary spaces that service Colombian cuisine

Colombian cuisine can be found in three different settings: fine-dining restaurants; comfort-food restaurants; and markets and market places. In all three, local food is consumed and its taste becomes a shared value among

Colombians. *Fine-dining restaurants* have become a space in which to socialize, to educate one's palate, and to offer new experiences to local consumers as well as to national and foreign tourists. For restaurateurs and chefs, these spaces are first and foremost a business opportunity, but they also serve as vehicles for the professionalization of the culinary craft, or for the transmission of the very same values that such spaces generate and which represent underlying community values (Chrzan 2006; Long 2004; Urry 2002; West and Carrier's 2004).[6] Even if many of these restaurants offer foods and dishes from other parts of the world, those which offer local dishes have become an important showcase for the different trends and manifestations of local cuisines (Toronja and Etiqueta Negra 2009; Wilk 2006).

The list below features several examples of dishes offered in the set menu of a couple of fine-dining restaurants in Bogotá:

- cream of beetroot soup, pork and beef meatballs with *paipa* (farm) cheese sauce, and baked potatoes (an *altiplano cundiboyacense*, or daily special) (Restaurant Salvo Patria 2015);

- starters: smoked cassava with elderberries and sour cream ("*costeño* style"); breaded squid simmered in hot chili oil with sprouted peas, lemon, and green mango; main courses: white bean salad, avocado, pickled green onions, and artichokes; pork *bondiola* (loin) braised with brown sugar paste, mashed creole potatoes, and cherry tomato salad; desserts: millefeuille *Patria* (topped with caramel spread and vanilla); tamarillo (or tree tomato) and blackberries with quinoa crumbs and ice cream (Restaurant Salvo Patria 2015);

- starters: thin slices of guatila (*citron*) or *papa de pobre* ("pauper's meal") with pickled *cubios*, *hibias*, and *chuguas* (Andean tubers); main courses: colored quinoa and rice salad with heirloom tomatoes and fresh cheese; smoked chicken with *tucupí* (a spice obtained from mandioca or cassava); *pusandao* (casserole of plantain and beef from the Pacific basin); dessert: maize tart with red berry sauce. (Restaurante El Panóptico, Vive; from personal archives.)

In turn, *comfort-food restaurants* have a much longer tradition in Colombia, and were in particular demand when cities expanded and workers were forced to spend more time at work. Even if their homes were relatively near their jobs, they did not have the time to go home and prepare lunch. This type of restaurant is aimed at this type of customer and serves typical home-style lunches, traditionally prepared with local ingredients. They are usually called *almuerzos ejecutivos* ("executive lunches"), *corrientazos* ("run-of-the-mill"), or *caseritos* ("mothers' cooking").

Some of these restaurants have adapted to their clients' daily routines and thus adopted contemporary trends and tastes, such as offering smaller quantities and better balanced diets, for example by increasing the amounts of greens and reducing carbohydrates, and employing fewer fats, thickening agents, and sweeteners. Nevertheless, their range is broad, and the actual content and price scale vary according to their target customer. In these settings cooks are not overly concerned by the concept of professionalization, but they must nonetheless follow well-established culinary techniques and possess a sound culinary knowledge. Their main concern is to preserve traditional knowledge and faithfully reproduce home cooking.

Examples of typical comfort food restaurant's menus for the day (*almuerzo del día*) are:

- starter: rice soup; main course: grilled steak, fried *yuca* (cassava), white rice with spinach, beet and carrot salad, guava juice; and dessert: fresh cheese with berry compote (Cositas Ricas, Bogotá);

- starter: cream of pumpkin or maize soup; main course: chicken in coconut sauce with sweet plantain slices, plain white rice and green salad, and pineapple juice (Andante Café, Bogotá).

Many *marketplaces* and *rural open markets* are also, by their nature, a way for people to see a daily display of local cuisines. In these spaces, the relationship established with the local produce is direct and cooks serve and prepare the dishes with the products they sell. It is in these venues that we can perhaps best see the full array of a country's regional cuisines, and the strongest expression of peasant foods and tastes—also known as *cocina popular* (local traditional cuisine).[7]

Some typical market dishes include: *empanadas* (stuffed pastries; liver and onions; maize soup; *cocido boyacense* (a stew from the Colombian High Andean plateau); *ajiaco santafereño* (a potato cream soup popular in Bogotá); cornmeal porridge (a typical soup from the Santander department); potatoes with peanuts (typical of Cauca region); flank steak in sauce *Criolla* (tomato, onion and garlic), plain white rice, *papas chorreadas* (potatoes covered with a tomato, onion, and cheese sauce) and avocado salad; *viudo* (fish stew), fish soup, shrimp ceviche; *tamales*; *arepas* (corn cakes); fruit salad and fruit juices (Personal archives; IPES, n.d.).

Techniques and meanings

I understand cuisine as a construct resulting from both knowing how to prepare and use certain foods (a technique) and the logic under which this knowledge is applied (a technology). Therefore, I focus on the culinary knowledge that these three settings generate about recipes and culinary techniques, but also on the individual motivations, social codes, and ethical values involved in cooking. Transmitted and explained by education and training, these techniques are the material with which daily practices create tradition and reproduce sets of social forms (Mauss 2007). Food habits and cuisine need to be understood as a human technique, a "knowing of how to do something" that is naturalized in social life as part of a particular symbolic order (Barthes 1975: 59). The argument made by Colombian anthropologist and chef Julián Estrada gives substance to this idea, especially when he affirms that each regional cuisine is different and unique, and should be understood as "one in which local technology uses accessories and materials that belong to the habitat, with almost exclusive products and spices, and one in which elements like air, the weather and fire mix with distinct flavors and the specialized hand to obtain a recipe that only in that place is originally prepared" (Estrada 2005: 215).

What I herein refer to as "culinary knowledge" is a composite of techniques and technologies that change from place to place and that in turn account for the actual differences between one cuisine and the other. In the case of fine-dining restaurants, professionalization is a necessary element in order to understand their particular culinary knowledge. Their trade is based on an education of marked European influence (what, in the trade, is known as *cocina internacional*), which in turn has played a significant role in teaching methods of technical and professional cuisine in public entities such as the Servicio Nacional de Aprendizaje (SENA) and its Centro Nacional de Hotelería, Turismo y Alimentos, founded well over forty years ago (SENA 2015). Since then, various culinary training courses offered by private schools and institutions have also disseminated these European-influenced culinary values and technical practices.

At the same time, in these same training centers, Colombian cuisine is increasingly taught to the same degree as the sociopolitical and cultural value of local cuisines increase. Furthermore, when high-end restaurants use local ingredients, they are compelled to draw on Colombian regional traditional culinary techniques and knowledge. Thus, international techniques are applied to local ingredients, local techniques to foreign ingredients, and local techniques to local ingredients. For example, Leonor Espinosa, one of the most acclaimed Colombian chefs today, is quoted as saying about her most famous restaurant: "[This cuisine is] product of the recreation of promissory spices brought to life in seas, rivers, mountains, prairies, valleys, jungles, and deserts, combined with our ancient culinary wisdom and in harmony with prevailing contemporary culinary philosophies" (Restaurante Leo Cocina y Cava 2015). Another well-known cook and food researcher from the Caribbean coast, Alex Quessep, when referring to what he thinks he expresses as a Colombian when cooking says:

> As time goes by, we add to our mix and mixtures, but rather than strictly acting on reason, we try to improve our awareness [. . .], yes, we are now more aware of that. Nevertheless, I'd like to say that we are now also less inhibited, less orthodox when mixing [. . .] We are no longer looking for somebody's approval or following any particular line, I think our cuisine rather responds to certain cultural, emotional aspects that relate to what we think we are.
>
> Personal communication, August 3, 2014[8]

In the case of comfort-food restaurants, both the techniques and most of the ingredients are local. Imported ingredients are unusual, but exceptions include the processed seasonings and sauces that have already become part of the Colombian diet, and meats, tubers, cereals, vegetables, grains, and other staple goods that were introduced into the traditional diet several centuries ago. Meals are prepared following traditional "home-cooking" techniques (to some extent influenced by European recipes and techniques, which for many years have been part of the national domestic repertoire).[9] Comfort-food restaurants usually offer a warm, homely (if somewhat formal), pleasant, and efficient experience. The general idea behind is to make people feel at ease.

Something similar takes place at the open market tables, but the setting, unlike that of comfort-food establishments, is more relaxed. In general terms, when dealing with traditional Colombian dishes—much more so vis-à-vis rural peasant cuisine—both techniques and ingredients are necessarily local (IPES, n.d.). The logic of cooking within these spaces is one-of-a-kind: to offer plentiful and generous portions of "home-cooked" foods. The service is guided by the same warm and efficient criterion followed at comfort-food restaurants, if in a rather more rustic ambience. A family-friendly atmosphere usually surrounds these tables, placed in such proximity to stalls selling vegetables, fruits, cereals, meats, and dairy products, that the connection to the land and the provenance of the produce is more evident.

A public/cultural/culinary perspective

The different interpretations of whatever is understood by the term "traditional" in these and other settings (that is, street-food vendors, catering for collective events, home-delivery services, specialty shops, and celebrations at home) have created an intricate culinary network with multiple edges, sociocultural planes, and vectors. But the network, even if unfinished and in some ways apparently incongruous, exists thanks to the presence of multiple principles and intentions directed towards one end: forging a sense of belonging and creating an identity in the midst of the dynamic process which cultural production implies (MacCannell 2008; Martín-Barbero 2002; Munasinghe 2001).

In such a context, it is said that the state has a duty to watch and protect both the conditions and the practices created by a community. Some of the tools governments have to perform these duties are public rules, guidelines of action, and strategies for the implementation of those prescriptions. A challenging question, however, in discussing the role of the state in this regard, is who are the subjects of state laws and how do these stipulations operate in daily life? We must also take into account the conditions that help or distract government agents from achieving an effective connection between formal regulations and real people with actual needs (Schwegler 2008; Wedel et al. 2005: 37). In this chapter I focus on the Public Policy for the Knowledge, Safeguarding and Promotion of Food and Traditional Cuisines of Colombia, adopted in February 2012 by the Colombian Ministry of Culture through the Division of Cultural Heritage, the institution in charge of establishing the guidelines for action with regard to cultural issues and following up their implementation.

According to the published document, this policy synthesizes the conclusions reached at seminars, conferences, meetings, and events held to discuss the urgent need to establish formal guidelines for regulating Colombian foodways. It also summarizes the information obtained by consulting experts and researchers, the analysis of several academic works on food studies, and the experience of governmental institutions such as the Ministry of Industry and Commerce (through its Vice-ministry of Tourism) and the Ministry of Environmental Affairs and Sustainable Development (Ministerio de Cultura de Colombia 2012: 11).

For the first time, the policy focuses on the cultural aspects of food, and in so doing it opens up space for new discussions at the sociopolitical level, establishing a bridge between public prescriptions of action with the way in which people actually experience and respond to political discourses around food. It states that through culinary knowledge and culinary practices people create bonds to a community and a particular region, and that traditional ways of food production, preparation and consumption are an essential component— an "indelible mark," in fact—of the cultural identity of a social group and, therefore, of the nation's intangible cultural heritage (Ministerio de Cultura de Colombia 2012: 9–10, 21). According to the white paper, cuisines are a cultural fact, and form a living tradition in that they constantly recreate themselves by means of daily culinary practices, oral descriptions, and continuous experiences of food consumption (Ministerio de Cultura de Colombia 2012: 21).

The idea of artisanal cuisines reinforces this relationship. It is said that family food businesses, street-food vendors, and farmers' markets are a fundamental part of the "universe of traditional cuisine:" "Given their scale, techniques, and idiosyncrasy, they have been called *cocinas artesanales*, an ambiguous concept which nevertheless does highlight the importance of 'making' and of 'doing,' and understands food as personal labor as well as a particular organization which merges cooks and family with apprentices" (Ministerio de Cultura de Colombia 2012: 25). Although rurality is seen as an essential feature of traditional cuisines, urban life also helps to define the "traditional", particularly so in the case of street-food vendors and farmers' markets that represent what Colombians usually understand as rural, thus tightening the imagined link between tradition and the countryside (Ministerio de Cultura de Colombia 2012: 38–39).

Another key point presented in this policy is that regional culinary traditions are poorly valued because they are usually seen as "rudimentary", as opposed to other foods that are seen as "prestigious" (Ministerio de Cultura de Colombia 2012: 65). Subsequently, one of the guideline principles states that, to safeguard the culinary heritage of Colombia, it is necessary to advocate for the recognition of cultural diversity and to raise awareness of the need to revive the sense of pride for local values (at a national, regional, and communal level) (Ministerio de Cultura de Colombia 2012: 74).[10] Beyond the value and novelty of this perspective, a topic of debate in such an institutional discourse is that it generates a space to potentially reify and exoticize local knowledge, based on assumptions about tradition and culinary authenticity (Camacho 2014. Also see for cases in other contexts Leitch 2003; Lewis 1989; Lu and Fine 1995).[11]

Projecting the local, interpreting the traditional

Both within and without the foodways context, we can understand "the local" as a physical and geographical territory, but also as culture; as that imponderable element that we experience as our common heritage. The sense of belonging to, of being part of a particular culinary context, and one's own identification with certain dietary habits, is predicated on the sense of being part of a territory, and of sharing values that such "modes of being" imply. In this sense, that which is "local" comprises tastes, habits, knowledge, appetites, and expectations. Colombian cuisine, whether "new," "traditional," "fusion," or "indigenous," responds to this double sense of space. Everyone from avant-garde chefs to "professional" and "traditional" cooks, restaurateurs, and entrepreneurs thrive under this umbrella that covers time and space, not only in the places I have described, but in the myriad interpretations that each diner must make when facing any of the many possible traditional cuisines.

At the very least, this is one of the ways in which traditions are differently interpreted in terms of techniques and technologies. Whether cuisines are reproduced by word-of-mouth from family to family, or from community to community, closely following both content and form in culinary practice, in one way or another content or form (or both) will be altered. Making sense of a cultural habit can occur through a significant act of reproduction and repetition—that is why crafts and handicrafts have an intrinsic value—but also through different ways of explaining or translating that habit from different perspectives (Hobsbawm and Ranger 1983). This is relevant for the purpose of understanding the complex dynamic processes underlying the imitation of foreign models and the recovery of local traditions when constructing new culinary identities in Colombia. Furthermore, the struggles of Colombians to define culinary expressions from the early twentieth century onwards is different from a past cultural identity but also different from foreign paradigms that lay, perhaps, in what Massey calls a "dislocation between the past and the present" (Massey 1995: 182). This is especially so when we think of the past as embodying the real character of a place. Such a dislocation could be a failure to recognize the preexistence of hybridity, in particular in regions with a strong colonial influence and where discourses about indigenous people have always been a matter of political and cultural struggle as well as a source of social anxiety.

One might say—as the three culinary settings that I discuss here show—that there is a vested interest in exaggerating the indigenous and the autochthonous, and in emphasizing a rich local diversity as a way of responding to the speed and potential anonymity of a globalized world. In this sense, Colombian elites and the ascending middle class exoticize themselves seeking to construct a local identity defined by its tolerance and respect of multiculturalism. At the same time, they enact cultural practices—such as the construction of a haute cuisine—which is supported by social differentiation and the commodification of indigenous culinary traditions. Also, although these discourses are produced by these privileged sectors of Colombian society, farmers, peasants, and indigenous groups can also exoticize themselves in their search for new strategies that will support their sense of regional and national belonging. In this way, they respond to, and partially determine, the course of their own sociopolitical reality.[12]

Disarticulation, allocation, and belonging

To conclude, I find it helpful to frame once again the relation between culinary knowledge and senses of belonging or estrangement that are enacted in the settings I have described—including the institutional perspective—with some reflections about the formation of collective identities and "culinary ideologies." Eduardo Restrepo and Santiago Castro-Gómez argue that "the national is constituted as a field of power from which different entities–identities are defined, normalized, and contested, also showing the predominant place that expert knowledge occupies within their categories and procedures" (Castro-Gómez and Restrepo 2008: 11–12). From a similar perspective, anthropologist Esther Sánchez—like many Colombian chefs, food researchers, and activists—declares herself highly committed to the revitalization of Latin American gastronomies, to resisting processes of cultural homogenization, and to consolidating differentiation through cultural expressions such as local, regional, and national cuisines (Sánchez, n.d.). This way of thinking implies that the authentic, in different spatiotemporal contexts, finds expression in products that represent it but do not possess it (Sánchez, n.d.: 12). In a way, in the process of making a Colombian cuisine, chefs and other actors involved seem to be compelled to exoticize the ingredients, techniques, and indeed the whole setting—including themselves—in order to make it legitimately "authentic" and to propagate it successfully.[13]

At the same time, however, they create a space for self-recognition and for sensorial and emotional experiences (Boden and Williams 2002; Gronow 2003). These experiences, as well as the ones lived in the process of creating and transforming local culinary forms, represent various ways of enacting the processes of modernization, globalization, and commodification (Martín-Barbero 2002; Green 2007). The development of culinary knowledge occurs through practices that reflect a rich, multilayered process of cultural production. It is our duty to examine and study from different perspectives and fields of knowledge what Colombian cuisine entails, what it consists of, which values it promotes, and up to what point, at this juncture, Colombians are adding (or not) their efforts to build a sense of belonging and empathy, and to enhance their feelings of "deep-rootedness," in addition to the simultaneous construction of social networks and exciting regional and national visionary projects.

Notes

1 A soup containing sundry types of meats, tubers, vegetables, and seasonings typical of several Latin American countries.

2 I use Priscilla Ferguson's understanding of the term cuisine as "the properly cultural construct that systematizes culinary practices and transmutes the spontaneous culinary gesture into a stable cultural code. Cuisine, like dining, turns the private into the public, the singular into the collective, the material into the cultural" (Ferguson 2004: 3).

3 According to the Departamento Administrativo Nacional de Estadística de Colombia (DANE), the revenues of the restaurant and cafeteria subsector grew 8.6 percent between October 2012 to September 2013, and again from October 2013 to September 2014; annual figures have been similar for the

past six years, in fact (DANE 2015a, 2015b.). On the other hand, according to the 2010 *Euromonitor International Consumer Food Service Report* for Colombia, "Quality is improving with regard to innovation and ingredient combination [. . .] in a country in which the popularity of eating out is increasing for both pleasure and convenience related reasons" (Euromonitor International 2010a). Another study made by Euromonitor International states that "the country's entry into the modern, globalized world has fostered a cultural change in the attitudes that Colombians have toward foods" (Euromonitor International 2010b). Besides the increasing opening of restaurants focusing on international gastronomies, "a new trend called 'fusion cuisine,' in which traditional Colombian and Latin flavors are mixed with international recipes and ingredients, has become widely popular" (Euromonitor International 2010c).

4   Unless otherwise stated, all translations are mine.

5   Although similar patterns are evolving across cities of all sizes in Colombia, I focus primarily on the capital city of Bogotá.

6   In her historical study of the invention and the development of the restaurant, Rebecca Spang (2000) examines social practices that have been, and continue to be, specific to restaurants and to gastronomic sensibility throughout the history of this public institution, even when it reflects larger socioeconomic and political orders. In this context, she notes how different moral and social values such as decency, honor, honesty, fairness, conviviality. and enjoyment were associated with the restaurant space and, more importantly, how the specifics of this association shift over time as the restaurant and gastronomic culture develop (Spang 2000: 149–150).

7   Bogotá has sixty-four market places of which nineteen are state-run public markets and the rest "community public" markets. Most of them make tables available to dine (IPES 2015).

8   I have in my possession an informed consent letter signed by the interviewee in which he authorizes me to disclose his actual name.

9   A few examples: hors d'oeuvre, rice pudding pasties (a typical Spanish dessert), boronia (plantain and egg plant puree), curry rice, roast beef in mustard sauce, roast beef simmered in cherry, macaroni with chicken, curuba mousse, marshmallow and coconut cake, bread and peach pudding (Instituto Colombiano de Cultura 1991; Nestlé 1988; de Zurek 1973).

10 It is beyond the scope of this chapter to discuss the extent to which this goal is related to nationalistic discourses. However, the sense of pride in local cuisines and local values is a key element in both the document cited and the different ways in which public and private agents and organizations approach the issue of food production and consumption in Colombia.

11 Patricia Aguirre's definition of what a policy should be resonates with this debate: a coherent corpus of explicit ideas that are the result of a comprehensive diagnosis that articulates actions with concrete goals and terms, in order to respond to a problem that has been already recognized by society itself (Aguirre 2005: 226. See also Wedel et al. 2005). For a discussion of romantic commodification, see also Long (2004) and Varul (2008).

12 For an interesting discussion of this phenomenon that Povinelli calls "optimistic multiculturalism," see Povinelli (2002).

13 For a rich discussion about the risk of homogenization and reductionism that the construction of a national (exoticized) cuisine entails, see Appadurai (1986, 2008), Ayora-Diaz (2012), and Bestor (2004).

References

Aguirre, P. (2005), *Estrategias de consumo: qué comen los argentinos que comen*. Buenos Aires: Miño y Dávila Editores.

Appadurai, A. (1986), "On Culinary Authenticity." *Anthropology Today* 2(4):25.

Appadurai, A. (2008), "How to Make a National Cuisine: Cookbooks in Contemporary India." Pp. 289–307 in C. Counihan and P. Van Esterik (eds), *Food and Culture: A Reader*. New York: Routledge.

Ayora-Diaz, S. I. (2012), *Foodscapes, Foodfields and Identities in Yucatán*. CEDLA: Amsterdam / New York: Berghahn.

Barthes, R. (1975), "Toward a Psychosociology of Contemporary Food Consumption." Pp. 47–59 in E. Foster and R. Foster (eds), *European Diet from Pre-Industrial To Modern Times*. New York: Harper and Row.

Bestor, T. C. (2004), *Tsukiji: The Fish Market at the Center of the World*. Berkeley: University of California Press.

Boden, S. and Williams, S. (2002), "Consumption and Emotion: The Romantic Ethic Revisited." *Sociology* 36(3):493–512.

Camacho, J. (2014), "Cómo se cocina una política pública de patrimonio culinario." Pp. 169–200 in M. Chaves, M. Montenegro, and M. Zambrano (eds), *Mercado, consumo y patrimonialización: Agentes sociales y expansión de las industrias culturales*. Bogotá: Icanh.

Castro-Gómez, S. and Restrepo, E. (2008), "Introducción: Colombianidad, población y diferencia." Pp. 10–14 in S. Castro-Gómez and E. Restrepo (eds), *Genealogías de la colombianidad: formaciones discursivas y tecnologías de gobierno en los siglos XIX y XX*. Bogotá: Editorial Pontificia Universidad Javeriana.

Chrzan, J. (2006), "Why Tuscany is the New Provence: Rituals of Sacred Self-Transformation Through Food Tourism, Imagined Traditions, and Performance of Class Identity." *Appetite* 47(3):388.

© Euromonitor International, Passport GMID database (2010a), Country Market Insight: Consumer Food Service Colombia (Nestlé Library, Cornell School of Hotel Administration).

© Euromonitor International, Passport GMID database (2010b), Attitudes to Eating Out. Electronic document, http://www.portal.euromonitor.com/Portal/ResultsList.aspx, accessed December 4, 2010.

© Euromonitor International, Passport GMID database (2010c), Eating Preferences. Electronic document, http://www.portal.euromonitor.com.proxy.library.cornell.edu/Portal/Magazines/Topic.aspx, accessed December 4, 2010.

DANE (2015a), http://www.dane.gov.co/files/investigaciones/boletines/mts/bol_mts_IIItrim_14.pdf (p. 6), accessed January 20, 2015.

DANE (2015b), http://www.dane.gov.co/index.php/es/servicios/muestra-trimestral-de-servicios-mts/82-economicas/comercio-y-servicios/5065-informacion-historica, accessed January 20, 2015.
Estrada, J. (2005), "Entre el fuego lento y el horno microondas." In *Desde los Andes al Mundo, Sabor y Saber. Primer Congreso para la Preservación y Difusión de las Cocinas Regionales de los Países Andinos*. Universidad de San Martín de Porres, Escuela de Turismo y Hotelería. Lima: Publicaciones Universidad San Martín de Porres.
Ferguson, P. P. (2004), *Accounting for Taste: the Triumph of French Cuisine*. Chicago: University of Chicago Press.
Green, G. L. (2007), "'Come to Life': Authenticity, Value, and the Carnival as Cultural Commodity in Trinidad and Tobago." *Identities* 14:208–224.
Gronow, J. (2003), *Caviar With Champagne: Common Luxury and the Ideals of the Good Life in Stalin's Russia*. Oxford and New York: Berg.
Hobsbawm, E. and Ranger, T. (eds) (1983), *The Invention of Tradition*. Cambridge: Cambridge University Press.
Instituto Colombiano de Cultura (1991), *Gran libro de la cocina colombiana*. Bogotá: Círculo de Lectores S.A.
IPES (2015), Electronic document http://www.ipes.gov.co/index.php/proyectos-institucionales/fortalecimiento-del-sistema-distrital-de-plazas-de-mercado, accessed January 20, 2015.
IPES (n.d), *Inventario de recetas propias e innovadas de las plazas distritales de mercado de Kennedy, La Perseverancia y Quirigua*.
Leitch, A. (2003), "Slow Food and the Politics of Pork Fat: Italian Food and European Identity." *Ethnos* 68(4):437–462.
Lewis, G. H. (1989), "The Maine Lobster as Regional Icon: Competing Images Over Time and Social Class." *Food and Foodways* 3(4):303–316.
Long, L. M. (ed.) (2004), *Culinary Tourism*. Lexington: University Press of Kentucky.
Lu, S. and Fine, G. A. (1995), "The Presentation of Ethnic Authenticity: Chinese Food as a Social Accomplishment." *Sociological Quarterly* 36(3):535–553.
MacCannell, D. (2008), "Why it Never Really Was About Authenticity." *Society* 45(4):334–337.
Martín-Barbero, J. (2002), "Identities: Traditions and New Communities.' *Media, Culture & Society* 24(5):621–641.
Massey, D. (1995), "Places and their past." *History Workshop Journal* 39(1):182–192.
Mauss, M. (2007 [1935]), "Techniques of the Body." Pp. 50–68 in M. Lock and J. Farquhar (eds), *Beyond the Body Proper: Reading the Anthropology of Material Life*. Durham, N.C. and London: Duke University Press.
Ministerio de Cultura de Colombia (2012), *Política para el conocimiento, la salvaguardia y el fomento de la alimentación y las cocinas tradicionales de Colombia*. Bogotá: Imprenta nacional.
Munasinghe, V. (2001), *Callaloo or Tossed Salad? East Indians and the Cultural Politics of Identity in Trinidad*. Ithaca: Cornell University Press.
Nestlé (1988), *Aliméntese bien en forma fácil y variada*. Bogotá: Nesté de Colombia S.A.
Povinelli, E. (2002), *The Cunning of Recognition: Indigenous Alterities and the Making of Australian Multiculturalism*. Durham: Duke University Press.

Restaurante Leo Cocina y Cava (2015), Electronic document: http://www. leococinaycava.com/, accessed January 20, 2015.
Restaurante Salvo Patria (2015), http://www.salvopatria.com/, accessed January 15, 2015.
Restaurante El Panóptico (2015), http://bogota.vive.in/restaurantes/bogota/ elpanptico/LUGAR-WEB-FICHA_LUGAR_VIVEIN-13118959.html, accessed January 15, 2015.
Sánchez, E. (n.d.), Potencial y riesgo de la gastronomía en América Latina. *I Congreso sobre patrimonio gastronómico y turismo cultural en América Latina y el Caribe, México*. Electronic document: www.conaculta.gob.mx/ turismocultural/ cuadernos/pdf/cuaderno1.pdf. Accessed November 30, 2010.
Schwegler, T. A. (2008), "Trading Up: Reflections on Power, Collaboration, and Ethnography in the Anthropology of Policy." *Anthropology in Action* 15(2): 10–25.
SENA (2015), Electronic document: http://senahoteleriaturismoalimentos. blogspot. com/p/blog-page.html accessed January 15, 2015.
Spang, R. L. (2000), *The Invention of the Restaurant. Paris and Modern Gastronomic Culture*. Cambridge, MA: Harvard University Press.
Toronja and Etiqueta Negra (eds) (2009), *Pescados Capitales: la vida en la orilla*. Lima: Editorial Toronja Comunicación Estratégica / Editorial Etiqueta Negra.
Urry, J. (2002), *The Tourist Gaze*. London: Sage Publications.
Varul, M. Z. (2008), "Consuming the Campesino: Fair Trade Marketing Between Recognition and Romantic Commodification." *Cultural Studies* 22(5):654–679.
Wedel J. R., Shore, C., Feldman, G., and Lathrop, S. (2005), "Toward an Anthropology of Public Policy." *Annals of the American Academy of Political and Social Science* Vol. 600:30–51.
Wilk, R. (2006), *Home Cooking in the Global Village: Caribbean Food from Buccaneers to Ecotourists*. New York/Oxford: Berg.
de Zurek, T. R. (1973), *Cartagena de Indias en la olla*. Bogotá: Editorial Bedout.

12

Cooking techniques as markers of identity and authenticity in Costa Rica's Afro-Caribbean foodways

Mona Nikolić, LAI, Freie Universität Berlin

When learning about the Costa Rican provinces in the third grade of elementary school, the main facts students are taught about Limón are as follows: Limón's capital is Puerto Limón, Costa Rica's international harbor on the Atlantic Coast; Limón is the province where the majority of Costa Rica's Afro-Caribbean population lives; and Limón is the province where rice and beans, *rondón,* and *pan bon* are eaten. While this can be seen as evidence of negligence regarding, and lack of interest in, Limón and its culture on a national level, it highlights the importance of the province's culinary culture as a marker of regional and cultural identity. The significance of this food culture is recognized as an identity marker by other Costa Ricans while at the same time members of the Afro-Caribbean population in Limón affirm their ability to cook Caribbean food "right" as a proof of their heritage.

Despite the relevance of this cuisine as an identity marker, in the last decades accounts describing Caribbean food culture in Costa Rica have noted its progressive disappearance from everyday life, even as the cuisine has become increasingly important in a tourism context. As international travel boomed from the late 1980s onward, so the demand for Caribbean food blossomed in popular tourist destinations. Indeed, this occcured in Puerto Viejo de Talamanca at a time when many local dishes—and the knowledge surrounding related preparation techniques—was on the verge of being forgotten. I will demonstrate how in this context—given the significance of authenticity as a measure for the legitimacy of a cultural experience—cooking techniques are gaining in significance, both as proof of authenticity of the Caribbean food on offer, and as a means of constructing and displaying local Caribbean identity. I argue that the importance of cooking techniques in tourist consumption contributes to their preservation and reintegration in the local context, at the same time that it facilitates the reinvention of local dishes. Before entering into the discussion of my fieldwork findings, I will briefly outline the theoretical framework on which my discussion is based.

Cooking techniques as embodied cultural capital

Pierre Bourdieu's theory of the habitus is the most prominent treatment of the relationship between consumption, the body, and society. According to Bourdieu, the habitus originates from the embodiment of the structures of the social field. These structures—the schemes of thought, perception, action, and evaluation—become manifest during the process of growing up in a particular social field, and become permanent bodily dispositions that govern the actions of the person without him or her being conscious of it (Bourdieu 2008: 100–105). The social field is defined through the struggles of different actors vying for the resources at stake. Here, different forms of "capital" play an important role.

For the purposes of this chapter, in discussing Bourdieu's concept of "capital," I consider cooking techniques as embodied cultural capital. Bourdieu defines capital as work accumulated either in material or in embodied form (Bourdieu 1983: 183). He distinguishes between different forms of capital among which he considers economic, social, and cultural capital as the most important. Like any form of capital, cultural capital can be converted, under certain circumstances, into economic capital and thus become institutionalized (Bourdieu 1983: 185). According to Bourdieu, cultural capital exists in three different states:

1 the *objectified* form, as in cultural goods, books, paintings, or machines;
2 the *institutionalized* form, as a form of objectification, as in academic titles;
3 and the *embodied* form.

This last form of cultural capital applies to knowledge that has become a permanent bodily disposition (Bourdieu 1983: 185). Cultural capital, through a process of embodiment, becomes "fixed" to the body: instead of being something possessed by the person who has embodied it, it becomes part of the person. Thus, embodied cultural capital cannot easily be transferred from one person to another and its accumulation is a process that occurs over time (Bourdieu 1983: 186–187).

This characteristic of embodied cultural capital—being fixed to the body of the actor—is most obvious in the case of practical forms of knowledge, such as cooking techniques. Practical knowledge is accumulated and embodied through watching and imitating the actions of others. Instead of being easily reproducible or transmittable through words, it is reproduced in actions. Actors are often oblivious of the processes by which practical knowledge is acquired, transmitted, and eventually naturalized. Cooking techniques as practical knowledge become not only part of our personal characteristics but also markers of identity, and in this latter regard can also be actively used to construct identities, as we see in the context of tourist consumption.

Tourism, authenticity, and cooking techniques

Since the 1970s, when Dean MacCannell introduced the concept of "authenticity" in the field of tourism studies, its quest has been considered one of the core motivations for tourists. MacCannell argued that tourists are concerned with experiencing "life as it really is lived" (MacCannell 1976: 94), that is, they are in pursuit of the "authentic" and the "real." Ning Wang points out that in tourism studies the concept is often used in a sense originally derived from the context of museology, where the main criterion in establishing the "genuineness" of an object is that it was made by local actors and produced according to local traditions (Wang 1999: 350–351). With regard to the context of tourist consumption and the role of cooking techniques, this would mean that only if the food were cooked by a member of the local community and prepared by using the local preparation techniques would it be considered "truthful." Thus, in the context of tourist consumption, cooking techniques play a central part as guarantees of authenticity.

In his theory of the "tourist gaze," John Urry stresses another aspect that should be considered when the concept of "authenticity" is applied in the context of tourist consumption. Urry defines the "tourist gaze" as a specific way of looking at things that is oriented towards perceiving the new, the unique, or the different. He distinguishes a "collective tourist gaze" from a "romantic tourist gaze," arguing that tourists with the latter search for the individual experience of "authentic" local cultures (Urry 2002: 40–45).

According to Urry, however, tourists judge a local culture as "authentic" only when it corresponds to their preconceptions of it. So it is only when a cultural performance accords with their expectations, heightening their "tourist gaze," that their experience is deemed "truthful" and "valuable."

In a similar vein, for local cuisine—as well as its preparation and presentation—to be accepted as "authentic" by international visitors, it has to match what they expect to find, as has been shown in the context of "ethnic" restaurants. As David Bell and Gill Valentine state, drawing from Erving Goffman, restaurants are contexts in which cultural and social norms and different understandings of a particular culture are negotiated (Bell and Valentine 1997: 125–126; also Long 2004: 38). In her study on "authenticity" and ethnicity in the Mexican restaurant industry, Marie Sarita Gaytán stresses how important it is for restaurants to live up to tourists' expectations, emphasizing how in these establishments the atmosphere, internal decoration, and employees' appearance play a part in creating "authenticity" (Gaytán 2008: 320–322).

Consequently, in the context of tourist consumption, "authenticity" is achieved by the fact that a local cook prepares a dish according to local custom, and also because the local cook prepares the food that tourists are expecting to find. Consequently, he or she must prepare it displaying the techniques that, from the tourists' point of view, are emblematic of the local food culture. As a result, the commercialization of a local food culture may lead to the conservation of "traditional" food culture and cooking techniques, and to the emergence of a new "authentic" local food culture, as is the case in Puerto Viejo de Talamanca.

Caribbean food culture in Puerto Viejo de Talamanca

This Afro-Caribbean settlement in the Talamanca emerged during the first half of the nineteenth century, when Afro-Caribbeans from Bocas del Toro, Panama, Bluefields, and the Colombian island of San Andrés immigrated to the Talamancan Coast (Palmer 2005: 21, 73). For many years, due to a lack of infrastructure as well as political and social reasons, the region and its inhabitants remained isolated from the rest of Costa Rica. The local population—mainly Caribbean and indigenous people—were heavily dependent on subsistence farming and their trade relations were largely orientated toward Panama. In the 1980s, at a time when villages were still relatively isolated, the first Europeans and North Americans settled in Talamanca, attracted by the untouched natural beauty and eager to pursue an alternative lifestyle. Puerto Viejo de Talamanca, the village where I conducted fieldwork from March through May 2011, soon became one of the most popular tourist destinations on the Atlantic Coast. It has become especially popular among eco-tourists, particularly since 1989 when the NGO ATEC (*Asociación Talamanqueña de Ecoturismo y Conservación*) established an office there, and a number of wildlife conservation projects were launched (*Caribbean Way* 2011: 19–22).

The boom in tourism led to changes in the population, triggering the immigration of national and international outsiders into the region at the same time that the local population emigrated, largely due to a lack of working opportunities (Vandegrift 2007: 125–127). Furthermore, the increase in visitor numbers introduced new alternatives to local food culture, as supermarkets began to sell a wide range of different and imported products, and restaurants began to offer different ethnic cuisines, catering for multiple tastes. The local population was no longer restricted to grown products in the area, and instead became able to integrate items from different culinary cultures into their foodways, undermining the importance of local cuisine in everyday life. In the context of tourist consumption, however, there has been a growing demand for the Caribbean dishes that are typically recommended as one of the culinary highlights for those visiting Costa Rica (Costa Rica Tourism 2012).

Despite the fact that Puerto Viejo de Talamanca is today a culturally diverse village (Sitios de Costa Rica 2004), at the national level it is regarded as Afro-Caribbean. As such, in addition to being a popular eco-tourism destination, it is suggested as a destination for those longing to experience a "Caribbean lifestyle" in Costa Rica. Visitors from Costa Rica and beyond travel to Puerto Viejo expecting to find people living happily, worry- and stress-free, in harmony with nature. They also expect to find a Caribbean food culture. In the context of tourist consumption, a Caribbean "stage" is set in several ways. The restaurants are often arranged as brightly colored wooden houses built on stilts, displaying a veranda. Indoors they are decorated with the Jamaican flag, and pictures showing sea or beach motives in which Pan-African colors dominate. They play reggae music as a sonic reference to Caribbean culture. The atmosphere in these restaurants is important in staging and constructing an ethnic or cultural identity (Gaytán 2008), but equally vital, from the tourists' point of view, is the presence of dishes and flavors representative of local food culture. In the case of the Caribbean food culture of Limón, these include dishes such as rice and beans and *rondón*, and pastries like *patí* and *pan bon*. The coconut milk and the fiery Scotch Bonnet pepper included in their preparation are for tourists emblematic flavors of, and pervasive in, Caribbean food culture. In Puerto Viejo de Talamanca, the interest of outside visitors in the Caribbean food culture emerged at a time when it was about to vanish, its significance in daily life having otherwise dwindled gradually for some time.

Cooking techniques as markers of Caribbean identity

The results of changes in the lifestyle and foodways of the local Afro-Caribbean population—and their critique of them—were obvious in my conversations with Johnny.[1] Taking his favorite dish (the *rondón*) as an example, Johnny lamented local people's renunciation of their own food culture and lifestyle:

> Yeah, the real women . . . they do this job, the authentic Caribbean women do the *rondón*, yeah, only authentic Caribbean women do the *rondón*. Why? Number one: They don't know how to—the majority of the women don't know how to chip a coconut now. And the banana, it stains your finger, you don't want to stain your finger, they wanna be painting now. And really pretty, you know? "You mad? I just paint my nails today. Never."
>
> JOHNNY, April 2011

In addition to Johnny's unfavorable view of the lack of knowledge or willingness among the predominantly younger local population to prepare Caribbean dishes, his emphasis on the "real" and "authentic" Afro-Caribbean women knowing how to prepare the *rondón* shows the interconnectedness of preparation techniques and local identity. From his point of view, those who have given up on their native cuisine and preparation techniques have also given up on their Caribbean identity: They are not "real" Afro-Caribbeans anymore.

As a marker of this identity, my interviewees all felt that the ability to cook Caribbean food is a trait inherent in the Afro-Caribbean population. When I asked Linda, a local chef and restaurant owner, who had taught her to cook, she told me: "Well, I don't [know] really—it's natural, it's part of our culture. I don't (sic) really learn it. You grow up watching people around you cook, putting things together. You just know it. It comes natural [sic] to you, you feel it" (Linda, April 2011). The ability to cook Caribbean food is not something that is explicitly taught, but it remains part of the Caribbean culture. By growing up in an Afro-Caribbean community and watching others cook, individuals embody cooking techniques and develop a sense of the "right" way to prepare and season, the "right" measurements, and the "right" taste. Both the act of learning to cook and cooking are unreflective processes. My interviewees repeatedly claimed to be cooking intuitively and did not believe that cooking Caribbean food could be learned by just passing on a recipe in written or oral form: "Well, people say, 'No, the thing is from your hand, that's why the things come out [right],' because people say 'You gave me the recipe, and I make it and it don't [sic] come out the same.' I think, 'Eliza, it's your hand'" (Eliza, August 2011). Eliza is expressing here the commonly shared view that being able to cook well, and especially to cook Caribbean food "right" is an embodied cultural knowledge that is considered a natural virtue in a person.

Taking up Bourdieu's theory of capitals outlined above, practical knowledge is unreflectively learned and embodied by growing up within the social field and by watching and imitating other members in the field. It becomes part of oneself. As embodied cultural capital, the knowledge of cooking techniques cannot be transmitted through speech, and can be passed on only through practice (Bourdieu 2008: 134–136). The ability to cook Caribbean food "right"— and "right" in this context also means "using 'traditional' cooking techniques"— is thus tied to being Caribbean. But besides marking Afro-Caribbean identity, cooking techniques are actively used to perform and construct an Afro-Caribbean identity in the context of tourist consumption.

Constructing and performing a Caribbean identity for tourist consumption: The *rondón* cooking class

In Puerto Viejo de Talamanca, the construction and performance of Afro-Caribbean identity takes place in the multiple settings where Caribbean food is commercialized (Nikolić 2014: 213–221). It is, however, most apparent in the context of Caribbean cooking classes. These classes are among the tourists' most requested activities with the local tour operator, ATEC. Visitors can choose between learning how to prepare rice and beans, preparing a vegetarian Caribbean meal, and (the most popular option) cooking the *rondón*. Johnny has been in charge of the *rondón* cooking class for several years, and enjoys putting to use knowledge about his culture and cooking techniques that he learned from his mother and grandmother. Tourists pay roughly US$25 to join the class, making it a valuable source of income for Johnny. I argue that his *rondón* cooking class is an example of how, in the context of tourist consumption, cooking techniques are used to perform and to construct an Afro-Caribbean identity, due to their function as markers of "authenticity."

The *rondón*, the local version of the Jamaican "rundown," is one of the dishes most representative of the Caribbean food culture for tourists. Because of its name (which in Jamaican English means "running down," or foraging for ingredients), descriptions of the *rondón* in tourist guidebooks, online, and also in scholarly studies, define it as a stew made from any ingredient that can be found in the woods, the sea, or on the farm, slowly cooked in coconut milk over a campfire. These sources describe people gathering around the campfire to share stories while the *rondón* cooks (Palmer 2005: 61; Ross de Cerdas 2002: 92–93; Costa Rica Tourism 2012). Thus, the preparation of *rondón* is perceived as being time- and effort-intensive but inexpensive. Another aspect that the description of its preparation makes apparent is the variability of the dish: It is made out of what one is able to gather locally.

In Limón, the *rondón*, and its preparation in particular, has become part of the Caribbean lifestyle romanticized by the tourist industry: a free and easy existence lived in harmony with nature. But at the same time that in the touristic context *rondón* is a dish emblematic of this perceived Caribbean way of life, it is also an extremely important dish in the social life of the local population; that is, it is essential at family celebrations. Yet despite this significance, most members of the local Afro-Caribbean community have stopped cooking it. From Johnny's point of view, the reason for this lack of interest in preparing the *rondón* is to be found in the laborious "traditional" techniques required to prepare the dish. He himself was very proud to be able to use them, and based much of his performance in his cooking classes on them.

Johnny's classes are held in his garden, where he has cleared a space to build a campfire and has surrounded it by a circle of benches made out of tree trunks. Right behind the fire he has placed a sink and a table. Before starting his classes, he places on this table all the ingredients he will use for his *rondón*, including coconuts, green plantains and bananas, cassava, yams, sweet potato, *malanga*, carrots, *tiquisque*, onions, red bell peppers, Scotch Bonnet peppers, thyme, celery, ginger, cilantro, garlic, red snapper, and, sometimes, even lobster. Another special ingredient in his *rondón* are garlic-flavored bouillon cubes, evidencing the integration of new ingredients not only into this specific dish but also generally into the local food culture.

His teaching is built upon the idea that the *rondón* is a "poor man's food," a perception that corresponds to the general image of the *rondón* as a dish made from foraged ingredients. Johnny considers the interest in getting to know "other people's" food and especially the "poor people's" food as one of the main reasons why tourists book his class.[2] Through his cookery performance he is living up to the tourists' expectation to know the "authentic" lifestyle of the less well-off. According to him, a crucial characteristic of "poor man's food" is that it is easy on the pocket but hard on the cook in terms of the labor required to make it:

> You wanna know why it's the poor man's food? All what you see we have there [sic], we could farm it. Provision it. So this is how we used to survive. So, the poor man's food is something you work for. It may be difficult, but you could make it. You'd never need to go to the supermarket to go and buy anything. You'll get from your garden and from the sea, you get your fish and you come back, you get your coconut, everything is there. You have to work hard.

JOHNNY, March 2011

In addition to upholding the image of *rondón* as a dish made from inexpensive, locally available ingredients and prepared using "traditional" cooking techniques and tools, preparing the *rondón* shows visitors how much time and effort is required. Thus knives or electric blenders are absent from Johnny's cooking class, and instead he relies on his machete for a variety of tasks, including breaking coconut shells, and peeling bananas and tubers. He also uses it to cut firewood, and while showing his clients how to do this—and letting some of them try for themselves—he explains how his grandfathers also used little other than a machete to prepare their meals. When he starts the fire, he begins to prepare the coconut milk, one of the most characteristic flavors of Caribbean food. He uses a coconut grater—a cooking utensil made out of a metal sheet that is perforated with a nail before being bent and attached to a wooden frame—to process the coconuts.[3] He soaks the coconut flakes in water for a while and then drains them to collect the indispensable coconut milk. He also invites his audience members to take part in this lengthy process, thereby allowing them to find out for themselves just how arduous it is. In this way, the use of the "traditional" cooking utensils and techniques becomes a way to entertain tourists.

During his classes Johnny also provides tourists with a sense of Caribbean culture, shaping his own performance to dovetail with the visitors' image of his food and culture. The cooking techniques and especially the application of a few simple utensils, as well as cooking on a campfire, reinforce the tourists' image of his lifestyle as "simple" and "in accordance with nature." Although Johnny claims *rondón* to be a "poor man's food," his definition contrasts with the significant amount of high-quality ingredients displayed on the table, and thereby enhances the romantic image of the "poor yet rich" life led by the local Caribbean population. Hence, his performance corresponds to the tourists' expectations and is therefore "authentic" from *their* point of view (Urry 2002: 59). On the other hand, it is an example of a local culture that has come into being in this particular way only through the impact of tourism.

When one compares Johnny's performance to the preparation of *rondón* in the local context, the fact that the dish is barely prepared these days by local Afro-Caribbean families is just one of a number of differences. Another major contrast lies in the way in which Johnny's version of the *rondón*, prepared from the wide range of ingredients noted above, is described as "poor man's food." Since the passing of the *ley de la conservación de la vida silvestre* (Asamblea Legislativa 1992. Ley N° 7317) and the privatization of land use (Palmer 2005: 256–257), it is no longer possible to prepare this dish from foraged ingredients; a trip to the supermarket is now necessary, rendering the *rondón* very expensive. In fact Johnny acknowledges these differences: "The poor man's food has become the rich man's food now" (Johnny, March 2011). He summed up this change in one interview, referring to the fact that *rondón* is now a dish popular among "rich" tourists, and also to the fact that *rondón* is now quite costly, even more so when one uses as many ingredients as he does. Today, he, like everyone else, buys his ingredients at the supermarket.

Instead of simply displaying the local food culture, Johnny is rather constructing a new "authentic" food culture that meets the tourists' expectations. In his performance the *rondón*, once an inexpensive dish made of whatever could be found locally, has become the symbol of a rich Caribbean culture and romanticized lifestyle. Furthermore, local Afro-Caribbean people have begun to view the new spin on an "authentic" Caribbean food culture as genuine. Johnny himself, even though he acknowledges the differences, does not consider his performance a misrepresentation of *rondón* and the local food culture, but rather the portrayal of a past local food culture and way of living to which it is desirable to return, at least as far as values such as "being independent from economic stresses" or "living close to nature" are concerned. He is trying to lead his life in accordance with this idea of local Caribbean culture: He has, for example, started to grow his own vegetables and fruit in his garden in order to be more independent. Although he sometimes makes coconut milk with an electric blender when cooking Caribbean dishes at home, and he does not believe that preparing it this way affects its flavor, he prefers to use the coconut grater and to cook on the campfire when preparing his own meals. He is also teaching his teenage son how to cook in this way and how to use the coconut grater, seeking to pass on his cultural heritage.

Explaining his preference for these techniques, Johnny stresses the time-consuming and arduous work required to prepare the dish and affirms this is the only "right" way to prepare a Caribbean dish. This illustrates his embodiment of schemes of thought, perception, and evaluation of the social field within which he operates (Bourdieu 2008: 100–105): Instead of simply using the cooking techniques to perform his Caribbean identity, his Caribbean identity is what leads him to prefer these techniques. While Johnny does not emphasize the flavor, other members of the local Caribbean population, commenting on Johnny's *rondón* preparation, point out the significance of using "traditional" preparation techniques to achieve an "authentic" taste. When explaining why he ordered his *rondón* from Johnny, Kelley, a member of the local Afro-Caribbean community simply stated: "A machine can't give it the flavor of the hand" (Kelley, March 2011). The "right" and "authentic" flavor of Caribbean dishes can be achieved only through manual labor and thus by preserving "traditional" preparation techniques. Johnny's preservation of these, and above all his preparation of the coconut milk using the coconut grater, is a marker of the quality and "authenticity" of the dish he prepares, both for the tourists and for local people. Members of the local Afro-Caribbean community praise Johnny's *rondón* as the "authentic" version of the dish. It is the ideal that other cooks preparing *rondón* have to match as far as quantity and type of ingredients, as well as taste, are concerned. This implies that coconut milk should be prepared manually, using the coconut grater. When the cook fails to prepare the dish according to these criteria, his or her dish is evaluated as not "right" and "inauthentic." Thus, cooking techniques signal both embodied knowledge and authenticity.

Looking at the changes in the list of ingredients for Johnny's *rondón* highlights how decisive cooking techniques are as markers of "authenticity." The fact that Johnny has integrated ingredients such as garlic-flavored bouillon cubes does not affect the overall perception of his *rondón* as "authentic." The definition of *rondón* as a dish that can contain any ingredient the cook is able to find makes the integration of foreign elements easier. Furthermore, this integration is facilitated by the emphasis placed on cooking techniques as markers of "authenticity" and the fact that ingredients such as coconut milk are representative of Afro-Caribbean culinary culture: As long as *rondón* is prepared using a coconut milk that has been prepared by hand with a coconut grater, the dish's "authenticity" is not questioned, regardless of the other ingredients included.

Concluding remarks

In this chapter I have focused on cooking techniques and their role as markers of "authenticity" and identity in the Afro-Caribbean community in Puerto Viejo de Talamanca in Costa Rica. I have shown that, at the local level, being able to prepare Caribbean food using local and traditional cooking techniques supports the Caribbean identity of the cook. However, in addition to being a mere marker of identity, cooking techniques can be actively employed in the performance and construction of identities, as is the case in the context of tourist consumption. Taking the *rondón* cooking class as an example I have outlined how cooking techniques in general, and the preparation of coconut milk in particular, are used to construct a Caribbean culture that meets the tourists' expectations and at the same time is considered "authentic" at the local level. On account of the emphasis placed at the local level on cooking techniques as markers of "authenticity," changes in recipes are made possible and do not affect the overall assessment of the dish as "authentically" Caribbean. The knowledge of cooking techniques—that is, embodied cultural capital—has turned into a source of income for local cooks when catering for both tourists and locals. Furthermore, there is now a growing awareness of the importance of the cooking techniques required to prepare "authentic" local dishes and an a growing interest in maintaining and transmitting that knowledge, at a time when it was about to vanish.

Notes

1 Johnny is in his thirties and works as a tourist guide. The real names of all interviewees have been replaced by pseudonyms. They all signed consent forms allowing me to change their names, and these forms are in my possession.
2 His understanding of the tourists' interests parallels prominent arguments in the context of research on tourism—the search of the tourist, that symbol of the "modern" person, for the natural and "authentic" way of life of the "others" (MacCannell 1976: 3; Urry 2002: 9) and of food tourists as especially eager to taste the food of "others."
3 The coconut grater, a symbol of the local Caribbean food culture, is now widely available in different sizes at souvenir stores.

References

Asamblea Legislativa de la República de Costa Rica (1992), *Ley de la Conservación de la Vida Silvestre*. Ley N° 7317.
Bell, D. and Valentine, G. (1997), *Consuming Geographies: We Are Where we Eat*. London: Routledge.
Bourdieu, P. (1983), "Ökonomisches Kapital, kulturelles Kapital, soziales Kapital." Pp. 183–198 in R. Kreckel, R. (ed.), *Soziale Ungleichheiten. (Soziale Welt Sonderband 2)*, Göttingen: Schwartz.
Bourdieu, P. (2008 [1987]) *Sozialer Sinn: Kritik der theoretischen Vernunft*. Frankfurt/Main: Suhrkamp Taschenbuch Verlag.
Caribbean Way (2011), "Puerto Viejo." 35:32–34.
Costa Rica Tourism (2012), "Southern Caribbean Coast Delights Foodies." http://www.tourism.co.cr/costa-rica-art-and-culture/costa-rica-food-and-cuisine/southern-caribbean-coast-delights-foodies.html. Accessed January 26, 2015.
Gaytán, M. S. (2008), "From Sombreros to Sincronizadas: Authenticity, Ethnicity, and the Mexican Restaurant Industry." *Journal of Contemporary Ethnography* 37:314–341.
Long, L. M. (2004), "A Folkloristic Perspective on Eating and Otherness." Pp. 20–50 in L, M. Long (ed.), *Culinary Tourism*. Lexington: University Press of Kentucky.
MacCannell, D. (1976), *The Tourist: A New Theory of Leisure Class*. London: Macmillan Press.
Nikolić, M. (2014), "Reinventing Local Food Culture in an Afro-Caribbean Community in Costa Rica." Pp. 201–223 in B. W. Beushausen, A. Commichau, P. Helber, and S. Kloß (eds), *Caribbean Food Cultures. Culinary Practices and Consumption in the Caribbean and its Diasporas*. Bielefeld: Transcript Verlag.
Palmer, P. (2005 [1979]), *What Happen: A Folk-History of Costa Rica's Talamanca Coast*. Miami: Distribuidores Zona Tropical, S.A.
Ross de Cerdas, M. (2002 [1991]), *La Magia de la Cocina Limonense: Rice and Beans and Calalú*. San José: Editorial de la Universidad de Costa Rica.

Sitios de Costa Rica. (2004), "Talamanca." http://www.sitiosdecostarica.com/cantones/ Limon/talamanca.htm. Accessed August 6, 2012.

Urry, J. (2002 [1990]), *The Tourist Gaze: Leisure and Travel in Contemporary Societies*. Newbury Park and London: Sage Publications.

Vandegrift, D. (2007), "Global Tourism and Citizenship Claims: Citizen-Subjects and the State in Costa Rica." *Race/Ethnicity. Multidisciplinary Global Perspectives* 1:121–143.

Wang, N. (1999), "Rethinking Authenticity in Tourism Experience." *Annals of Tourism Research* 26(1):349–370.

Afterword

Carole Counihan, Millersville University

This volume takes a revealing trip through the kitchens, tools, and techniques of diverse mestizo and indigenous Latin Americans in Mexico, Guatemala, Venezuela, Peru, Colombia, Brazil, Cuba, Costa Rica, and the United States. This sweeping overview embedded in specific case studies shows the wealth of insights generated by focusing on how people use material culture and corporeal skills to transform raw ingredients into culturally meaningful comestibles. Chapters demonstrate the value of anthropological approaches and make use of archaeology, ethnohistory, and ethnography to generate rich empirical data and thoughtful interpretations of diverse quotidian and ritual culinary practices. Other researchers can take inspiration from sources used here, which include historical description by early colonial chroniclers; dictionaries and cookbooks describing ingredients, tools, and methods; excavations and surveys of habitations, butchering sites, ovens, and hearths; zooarchaeological, paleobotanical, participant-observation, and interview data; and material culture inventory. Articles show a variety of sites where technologies can be studied—in indoor and outdoor home and community kitchens, street stalls, restaurants, stores, supermarkets, factories, the suitcases of migrants, and the mass media.

Several key themes emerge from the book and suggest avenues for future research. One is the interrelationship between taste, place, technology, and access. Tools and techniques are key forces in determining taste—along with ingredients, of course—and these are often tied to specific places, environments, and cultures. As migrants travel the globe, they sometimes cannot find their tools and technologies, or bring them with them, because they are bulky and heavy. Moreover, as cultures change, artisans who make culinary tools may dwindle, and raw materials such as wood or clay may be less available due to environmental degradation and rising costs. Some precious skills and tools may only be accessible through inheritance or apprenticeship, and thus are tied not only to the home place but also to specific and often elderly people there. Some technologies employed in kitchens in the home country move to the industrial sphere in the diaspora.

All these processes alter familiar tastes, as happened with the demise of traditional skills and tools and the industrial production of *tortillas*, chili powder, and canned *chili con carne* in the Mexican diaspora.

A second theme that emerges from the chapters is people's flexibility in using diverse techniques for different aims and in different contexts. As Fernández-Souza shows, the Maya use *metates* to grind not only foods but also ceramic materials and pigments; their tools have both quotidian and ritual functions; and they alternate seamlessly between the ancient and contemporary *metates*, metal hand mills, and electric blenders depending on time constraints, desired tastes, and ingredients. People adopt or resist new technologies for multiple reasons: cost, nostalgia, ritual connotations, cultural capital and distinction, or identity considerations.

A third theme of the book is how technology reflects and shapes class, gender, national, and ethnic identity. Brazil today is witnessing an upsurge in New Latin American cuisine in middle-class homes, where people express a high status identity by adopting new kitchen design, techniques, ingredients, cookbooks, and tools. In the Yucatán, people may choose "gastro-sedentism" and adhere to one culinary tradition, or "gastro-nomadism" and practice diverse cuisines using new tools, techniques, and cookbooks to demonstrate cosmopolitan identities. Some may reject traditional techniques because they symbolize subordinate status or a devalued indigenous identity or they may, like the Zapotecs of Tehuantepec, combine "modern" and "traditional" identities in their technologies and revalorize their "ethnic" food during religious feasts where iconic cultural meals take center stage.

A final theme I want to highlight is the oscillating demise and re-appropriation of traditional technologies. In today's world, "authentic" cuisines and their associated material culture are valued by powerful culture brokers in some settings, but despised or ignored in others. For example, indigenous techniques for manioc production in the Venezuelan Amazon are erased in the adoption of *casabe* bread in urban domestic kitchens and restaurants. But in Peru, foreign-trained chefs have re-appropriated historically devalued indigenous ingredients and developed a fusion of Andean and Amazonian cuisine identified as "native" and valued as symbolizing Peruvian cultural and biological diversity, launching a new gastronomy appealing to a global cosmopolitan culinary market. Colombian restaurant chefs' understanding of "local," "traditional," and "artisanal" cooking contributes to a new cuisine valued in the national Culinary Academy and shaping the curricula of culinary schools and university programs. In the Costa Rican Caribbean, traditional preparation techniques are fading from everyday life while at the same time they have become sources of cultural capital and markers of identity and culinary authenticity in the tourism industry.

The themes explored in this provocative book raise a broad array of questions for future research on culinary technology:

- Given that globalization, emigration, and industrialization produce transformations in cooking techniques and technology, and given the importance of cuisine in representing culture, how do migrants maintain their culinary identities in the face of homogenizing forces of industrial ingredients and processes?
- Who carries technologies and what sorts of social, economic, and political forces influence the value of traditional tools and skills and determine whether they will be retained or abandoned as new technologies become available?
- How do fusion cuisines depend on the blending of diverse ingredients, tools, and skills, and what effect do they have on local and transnational economies and identities?
- What roles do identity hierarchies and conceptions of authenticity play in the acceptance or rejection of new and old culinary technologies?
- How does the intertwining of technology, skill, taste, and emotion affect social change and the acceptance or rejection of globalized foodways and development projects?
- What impact do formal institutions such as culinary schools, television stations, hygiene regulators, or geographic indication legislation have in supporting or hindering new and old culinary technologies?

This book explores a most basic dimension of human experience—material need satisfaction—by examining the means of production in culinary practice, which are fundamental to the household economy and family survival. They are also intertwined with global economies through the movement of goods and peoples, through the mass media, and through social and political institutions. Thus cooking tools and techniques are important vehicles and symbols of cultural persistence and social change and important foci for future research.

Notes on the contributors

Steffan Igor Ayora-Diaz (PhD McGill University, 1993) has been Full Professor of Anthropology at the Universidad Autónoma de Yucatán since 2000 and currently directs a team project funded by CONACyT entitled "New Technologies and Contemporary Life in Urban Yucatán." He has published two monographs: *Foodscapes, Foodfields and Identities in Yucatán* (CEDLA and Berghahn 2012); and in Spanish *Local Healers and their Struggle for Recognition in Chiapas, Mexico* (Plaza & Valdés 2002). In addition, he has edited and co-edited four books in Spanish dealing with globalization, modernities, consumption, and cultural representation. He has published papers and book chapters on food and identity, taste, the anthropology of performance, Chiapas healers, and Sardinian pastoralism. He was President of the Society of Latin American and Caribbean Anthropology from 2011 to 2014.

Hortensia Caballero-Arias (PhD University of Arizona, 2003). Her main research areas are development anthropology, political and historical anthropology, ethnicity, and politics of identity, especially among indigenous peoples of the Venezuelan Amazon. She is Research Associate at the Center of Anthropology at the Instituto Venezolano de Investigación Científica (IVIC). Her most recent books are *Los Yanomami* (Caracas, Fundación Editorial El Perro y la Rana, 2012) and *Desencuentros y Encuentros en el Alto Orinoco. Incursiones en territorio Yanomami, Siglos XVIII–XIX* (Altos de Pipe, Venezuela, 2014). At present she is one of the coordinators of the Venezuelan Network of Interculturality and Legal Anthropology (REVIAJU).

Margarita Calleja Pinedo (PhD University of Texas at Austin, 2001). Professor at the University of Guadalajara, Mexico, at the Department of Regional Studies. Her fields of research are agro-food systems, the globalization of food, and food identity. Her most recent publication in Spanish, co-edited with Humberto Gonzalez, is *Agro-Food Territorial Dynamics in Times of Globalization* (Universidad de Guadalajara, 2004) (http://www.ciclosytendencias.com/vinculos/oportuno1.php).

NOTES ON THE CONTRIBUTORS

Carole Counihan (PhD University of Massachusetts, Amherst 1981) is Professor Emerita of Anthropology at Millersville University and studies food, culture, and gender in Italy and the United States. She is author of *A Tortilla Is Like Life: Food and Culture in the San Luis Valley of Colorado* (2009), *Around the Tuscan Table: Food, Family and Gender in Twentieth Century Florence* (2004), and *The Anthropology of Food and Body: Gender, Meaning, and Power* (1999). She is co-editor of several books including *Food and Culture: A Reader* (1997, 2008, 2013) and *Food Activism: Agency, Democracy, Economy* (2014). She is editor-in-chief of the scholarly journal *Food and Foodways* and has been conducting ethnographic research on food activism in Italy since 2009.

Juliana Duque-Mahecha (PhD candidate, Department of Anthropology, Cornell University). She has published papers and a book review in Colombian journals: "Cuatro plazas más cerca de la tierra," in *Cocina Semana* No.62 (publication pending. Bogotá, Colombia: Publicaciones Semana). "La cocina en Colombia, un desafío emocionante," in *Revista Semana tradición y cambio* (April 20, 2013. Bogotá, Colombia: Publicaciones Semana, pp. 30–31. "Los Ajiacos colombianos," (coauthored with Shawn Van Ausdal), in *Revista de Estudios Sociales* 29 (April, 2008. Bogotá, Colombia: Universidad de los Andes, pp. 158–165.) Review of Mary J. Weismantel, *Food, Gender and Poverty in the Ecuadorian Andes* (Prospect Heights: Waveland Press, 1988), in *Revista de Estudios Sociales* 29 (April, 2008. Bogotá, Colombia: Universidad de los Andes, pp.176–179).

Jane Fajans (PhD Stanford University, 1985) is a Professor of Anthropology at Cornell University. She has conducted extensive fieldwork with the Baining of East New Britain, Papua New Guinea from 1976 to the present, and on food and identity in Brazil from 2006 to the present. She is the author of *They Make Themselves: Work and Play among the Baining of Papua New Guinea* (University of Chicago Press, 1997), and *Brazilian Food: Race, Class, and Identity in Regional Cuisines* (Berg, 2012), and author and editor of *Exchanging Products: Producing Exchange,* Oceania Monograph 43. She has written on emotion and personhood and has an article on "Autonomy and Relatedness" in *Critique of Anthropology*. She is currently conducting research on regional foods and regional identity in Brazil and Portugal.

Lilia Fernández-Souza (PhD in Mesoamerican Studies, University of Hamburg, 2010) is Professor at the Universidad Autónoma de Yucatán. Her current research interests are focused on household, gender, and agency archaeology, archaeometry, Maya foodways, and ritual. She has edited two books in Spanish: *At the Jaguar's Ancient Kingdom* (Secretaría de Educación Pública, 2010) and *Everyday Life of the Ancient Maya in the North of the*

Peninsula of Yucatán (with R. Cobos, UADY, 2011). She has several papers and articles published and in press, both in Spanish and English.

Julian López García (PhD in Anthropology, Universidad Complutense, 1991) is Professor of Anthropology at the Universidad Nacional de Educación a Distancia (UNED), Spain. He has directed several projects in Guatemala, Bolivia, and Spain, focusing on the anthropology of food and the anthropology of violence and memory; He currently directs the "Indigenous Peoples and Modernity in Latin America" project, I+D+I (Research+Development+Innovation). He is the author, co-author, or editor of numerous books, papers, book chapters: some of his more recent work in Spanish includes *Dinosaur Reloaded: Contemporary Violence in Guatemala* (co-author; FLACSO, 2015) and *Kumix: Rain in Maya Myth and Ritual among the Ch'orti'* (author; Cholmasaj, Guatemala, 2010).

Claudia Rocío Magaña González (PhD Colegio de Michoacán, 2012) works at the Centro de Investigaciones en Comportamiento Alimentario y Nutrición (CICAN) at the Universidad de Guadalajara. Her main fields of research are: the anthropology of food, focusing on ethnic *fiestas* in Oaxaca; food sovereignty movements in Jalisco; and political and ethnic processes in México. Her most recent book in Spanish is *Alimentary Habits: Psychobiology and Socio-anthropology of Nutrition* (co-editor, McGraw-Hill 2014). She also has a book chapter in Spanish, "Understanding Obesity in Mexico: An approach from the Perspective on the Decolonization of Bodies," in *México Obeso* (Editorial Universitaria, Universidad de Guadalajara, 2014).

Lorenzo Mariano Juárez (PhD in Anthropology, UNED 2011 [Extraordinary PhD Award 2012]) is Assistant Professor at the Universidad de Extremadura, Spain. His research focuses on the anthropology of food and development, and medical anthropology and violence. He has been conducting research and fieldwork in Guatemala and Spain since 2004. His most recent publication in Spanish is *Hunger in the Ch'orti' Region in Easter Guatemala: Culture, Politics and Representation in the Dialogues on Cooperation and Development* (G9 Editorial, 2014). He has also published a number of papers and book chapters in Latin American and Spanish journals.

Raúl Matta (PhD in Sociology, University of Paris – Sorbonne Nouvelle [IHEAL], 2009) is research fellow at the University of Göttingen as part of the DFG Project "Food as Heritage." His research interests include food studies, the anthropology of cultural heritage, and urban sociology. Among his most recent publications are: "Valuing Native Eating: The Modern Roots of Peruvian Food Heritage" (*Anthropology of Food*, 2013); in Spanish, "Gastronomic

Republic and Cooks' Country: Food, Politics, Media and the New Idea of Nation in Perú" (*Revista Colombiana de Antropología*, 2014) and "Contrasting Faces in Perú's Patrimonialization of Food: Actors, Logics, Issues," In Ch-E. de Suremain and J-C. Galipaud (eds) The Fabric-actors of Heritage: Implication and Participation of Researchers in the Global South (IRD Editions and Editions de l'Etrave, 2015).

Mona Nikolić (PhD Institute for Latin American Studies, Freie Universität Berlin, 2014) In her recently published PhD thesis, *Identität in der Küche: Kulturelle Globalisierung und regionale Traditionen in Costa Rica* (Transcript, 2015) she examines the global-local relationship in the field of cuisine and eating habits as identity markers in Costa Rica. Her research interests include the anthropology of food and consumption, transnational studies and globalization, and the anthropology of Central America.

Ramona L. Pérez (PhD University of California, Riverside, 1997) is Professor of Anthropology, Director of the Center for Latin American Studies, and Director of the J. Keith Behner and Catherine M. Stiefel Program on Brazil at San Diego State University. She has worked on migration and health, identity and nationhood, and the formation of community among Oaxacan migrants in the US. Currently she works on bi-national youth identity and family composition, migration and identity, shifts in culinary food practices and nutrition, migrant youth in the context of deportation and survival, and the moral economy of lead poisoning in ceramic production. She has published in English and Spanish in the fields of anthropology, geography, public health, social work, criminal justice, and medicine. She currently serves on the Executive Board of the American Anthropological Association and was President of the Society for Latin American and Caribbean Anthropology from 2008 to 2011.

Anna Cristina Pertierra (PhD London University, 2007) is Senior Lecturer in Cultural and Social Analysis at the University of Western Sydney. Her research encompasses urban domestic life, material culture, consumption and the anthropology of media, especially television. She conducts ethnographic research in Cuba, Mexico, and the Philippines. Recent publications include *Locating Television: Zones of Consumption* (with Graeme Turner, Routledge 2013), *Consumer Culture in Latin America* (with John Sinclair, Palgrave 2012) and *Cuba: The Struggle for Consumption* (Caribbean Studies Press 2011).

Index

achiote (*bixa orellana*) 20, 21, 23, 56, 94
aesthetics (*see* food)
affect 61, 87
affective 94, 96, 97, 113
affectivity 65
    culinary
    cultural 38
Amazonas State 42
Amazon basin 43
Amazonian people 41, 47, 48
Andean
    cuisine 182
    food (*see* food)
    ingredients 150
    mushrooms 146
    tubers 142, 144, 145, 146, 148, 155
Arawak 44, 46
archaeology 181
artifacts 44, 49, 50, 72, 120
assemblage 97
    cooking 95
    culinary 95, 96, 97
    technological 87, 88
atol, atole 20, 63, 75, 76
attachment
    affective 7, 85, 86, 87, 93, 96
    culinary 95, 97
authentic
    cuisine 182
    local culture 170
authenticity 6, 7, 141, 168, 169, 177
    culinary 182, 183
    and national cuisine 8, 170
    re-creation of 8
    in tourism consumption 170
Austex 80–1

Batteau, A. W. 4
Bell, D. 170

blender 20, 59, 91, 92, 105, 112, 113, 130, 175, 176, 182
body
    and embodiment 169
    and soul 31
Bourdieu, P. 168–9, 173

cacao 20, 21, 23
capital 168
    culinary 117, 122
    cultural 168–9, 173, 177, 182
    symbolic 89
capsicum annum 73
    baccatum 144
Caribbean
    food culture 167, 171, 176
    identity 168, 173
carne con chile 7, 71, 73, 78, 81
casabe (cazabe) 42, 43
    preparation of 45–51
cassava 7, 41
    cooking stages 47–9
Castro, F. 132
Cerruti, M. 101
chapulines 102
chef 111, 116, 140, 155
    avant-garde 160
    celebrity 6, 8, 117, 121
    Colombian 157, 162
    elite 140
    local 139, 172
    Peruvian 140, 141, 147, 148, 150
    restaurant 117, 120, 182
chemical techniques 17
chile ancho 80
chili 18, 21, 56, 71, 73–7, 102, 144
    canned 94
    ground 79
    oil 155

INDEX

powder 79–80, 182
  sauce 21, 77–9, 94
chili con carne 7, 77–82, 182
Chili Queens 77, 81
chilhuacle 103, 106
chipil, chipilin 103
cocina popular 156
Cold War 127–8
comal 21, 22, 30, 32–6, 38, 77, 107
  clay 7, 24, 105
  iron 35–6, 39, 76, 99
comizcal 58–9, 61
cookbooks 5, 6, 21, 72–4, 78, 80, 82, 86, 89–91, 95, 97, 116, 120–1, 181, 182
  Peruvian 141
  regional 86
  stores 117
  as technology of memory 88
cooking 15, 21–2, 56, 72
  appliances 7, 88, 91, 133
  classes 116, 117, 122, 173–6
  domestic 58, 117
  fusion 154
  home-style 154, 156, 158
  implements 2
  ingredients 93, 94
  instruments 4, 7, 60, 63
  knowledge 72
  logic 158
  magazines 5, 74, 90, 117, 120, 154
  practices 1, 5, 16, 45, 61, 65, 133
  procedures 71, 72, 81
  process 32
  professionalization of 139, 140
  regional 86
  restaurant 141
  shows 117, 121
  skills 60, 77, 117
  sous-vide 149
  space 3, 18, 89
  technique/s 4, 8, 22, 50, 61, 87, 88, 92, 95, 141, 168–70, 172–3, 177, 183
  technology/ies 8, 16, 22, 87, 95, 140, 183
  tools 75
  tradition 88, 88
  traditional 35

utensils 75, 77
vacuum 149, 150
Yucatecan 85, 95
cooks
  domestic 59–60, 64, 115
  hired 113
  home 85–6, 91–2, 94, 96, 114, 160
  professional 160
  regional 96, 112
  restaurant 156
  traditional 55, 56, 60, 64, 160
Costa Rica 8, 167, 181, 182
cuisine 56, 72, 156–7, 159, 162 n 2
  Andean 182
  artisanal 159–60
  authentic 182
  Brazilian 111, 116
  Colombian 153–5, 157, 160, 162
  fusion 8, 140–1, 146, 163 n 3
  haute 5, 8, 111, 116, 161
  as human technique 157
  as identity marker 167
  Istmeño 7, 56–7, 62
  local 55, 96, 156, 170
  Mexican 106
  national 7, 8. 90, 115, 164 n 13
  native 154
  nouveaux 5, 140
  nouvelle 111
  Oaxacan 56, 100, 102–4
  peasant 158
  Peruvian 139, 140–1
  professional 157
  regional 55, 56, 62, 64–5, 85, 97, 100, 156, 157
  rural 24
  traditional 105, 154, 156, 160
culinary
  assemblage 95, 96, 97
  attachment 95
  authenticity 160, 182
  capital 117, 122
  change 38
  culture 141, 146, 167, 171
  exchange 77
  field 86
  heritage 81, 160
  identity 72, 161, 183
  industry 140, 154

knowledge 62, 72, 77, 90, 150, 157, 159, 161, 162
memory 5
modernity 36
patrimony 73
practice 4, 15, 36, 45, 58, 60, 62, 87, 88, 94, 104, 111, 159, 162 n 2, 181
realm 154
school 140, 148, 154, 182, 183
skills 117, 140, 143
system 44, 45
taste 94
techniques 95, 96, 150, 156, 157
technology 3, 15 31, 35, 38, 39, 101
tools 181
tradition 6, 15, 87, 93, 154, 160, 161, 182
triangle 2
values 2, 5, 55, 64, 144, 157
cultural capital 168–9, 182
  embodied cultural capital 168, 169, 173, 177
  production 96, 159, 162
  homogenization 162
curry 73, 74, 82, 163 n. 9
cuy (*see* guinea pig)

de Certeau, M. (with L. Giard and P. Mayol) 134
development
  agency 7
  initiative 127
  project 37, 40 n 6, 183
developmentalism, -ists 29, 35
dining out 115, 154
discourse 150
  aesthetic 154
  chef's 141
  culinary 140
  institutional 160
  political 159
domestic
  animals 18, 57
  appliances 3, 6, 81, 129, 132, 133, 134
  comfort 129
  consumption 133
  cooking 58
  cooks 86, 94, 115
  kitchen 88, 95, 182
  labor 112
  life 127, 130
  modernity 128
  practices 55, 88
  relations 42
  socialist modernity 127
  space 2, 127
  sphere 104
  structures 19
  technologies 78, 132
  tool 32
  work 4
  workers 112, 114, 117, 118
Dufour, D. L. 47, 51

eating
  culture 77
  guinea pigs (*cuy*) 144, 147
  habits
  out 115, 163 n 3
  pattern of 115
  space 18
emotion 100
  work 32
emotional
  experiences 162
  gap 36
  involvement 31
  value 31, 32
  wellbeing 133
Energy Revolution (Cuba) 126, 129, 130, 131, 132–4
ethnicity 55, 64, 65, 65 n 4, 170
  performance of 66
  Zapotec 56
ethno-archaeological 6, 15, 16, 17
ethnography 181
ethnohistory 181
Everton, M. 88
everyday
  consumption 126
  cooking 21, 61, 65, 94, 133
  culinary practices 4
  culinary techniques 3
  diet 35, 57, 93
  drudgery 2

INDEX

food 5, 23, 63, 93
household space 125
life 3, 85, 87, 134, 167, 175
meals 2, 5, 87, 88
objects 5
practices 5, 9, 55, 87, 134

Feeding American Project 71, 73, 74, 79, 80, 82
Fehérváry, K. 127–8, 134
feijoada 114, 121
food
  aesthetics 4, 141
  Andean 145, 149
  authentic 86, 162,
  Caribbean 167–8, 170, 171, 172–3, 175, 176, 177
  culture 71, 74–5, 167, 170–2, 174, 176
  ethnic 55, 106, 182
  fusion 2, 139
  gentrification 141
  homemade 71, 77
  identity (and) 15, 96, 101, 121
  indigenous 140, 142, 144
  industrially processed 5
  local food culture 141, 154, 170–1, 174, 176
  processed 87, 93
  system 42, 44–5
  taste (and) 1, 2, 5, 20, 50, 87, 94, 95, 97, 103, 105–7, 140, 146, 154, 156, 177
  Yucatecan 85
foodies 92, 140, 149, 150
foodways 75, 76, 159, 160, 172, 183

gastro-sedentism 182
gatrono-nomadism 182
Gaytán, M. S. 170
Gebhardt, W. 79–80
gender 182
  arenas 65
  differences 64
  distinctions 31
  politics 2, 3, 126
  practices 22
  roles 4, 16, 50

Giedion, S. 3–4
Gil, V. 170
globalization 3, 4, 162, 183
global-local 4, 7, 9, 85, 88, 96
Goffman, E. 170
grater 44, 49
  coconut 175, 176–7, 178 n 3
  manioc 48
  mechanical 48
  stone 48
  wood 48
griddle (*see comal*)
grinding stone (*see metate*) 17, 18, 19, 23
guinea pig (*cuy*) 2, 8, 141, 143–4, 147, 148–9, 150

habitus 168
  bodily 91
  female indigenous 50
hearth 15, 18, 50, 51, 75, 181
  cement 22
  three-stone 18, 21, 22

identity 9, 15, 24, 101, 102, 182
  Brazilian 113
  Caribbean 168, 172–3, 176, 177
  collective 86, 96, 97
  culinary 72
  cultural 141, 159, 161, 167, 171
  ethnic 31, 55, 56, 58, 64–5
  icons 118
  local 66 n 10, 161
  marker 121, 167, 169, 177, 182
  national 8
  performance of 173, 177
  regional 57, 167
  Yucatecan 86
industrial (industrialized) 72, 79, 81, 86
  chili con carne 81
  equipment 93
  food 24, 21, 71, 87, 93, 94, 95
  ingredients 94, 97, 183
  packaging 7
  producers 90
  production 182
ingredients 3, 15–16, 18, 45, 65, 71
  Amazonian 150

Andean 150
Brazilian 111, 116
in chili sauce 78
in curry 73–4
Istmeño 57
local 23, 118, 145, 155, 157, 158
locally produced 106
in *mole* 103
in *moqueca* 119
native 141, 145, 150
non-local 23
pre-packaged 5, 87
processed 8, 96
raw 181
in *rondón* 173–7
Spanish 56
as technology 86, 88, 93–5
as technological objects 5
traditional 2, 7, 24
Istmeño
cuisine 7, 56–7, 62
regional identity 57–8

kitchen 3–4, 58, 64, 88
appliance 39, 89, 126, 127, 130
community 56, 62–4
contemporary 1, 120
design 182
equipment 86, 115, 120
household/domestic 58–62, 95
indigenous 32
Latin American 1–2
as marginal space 2
modern 9, 60, 95, 129
politics 126–9
poyetón 38
renovated/upgraded 117, 132
rural 55–6, 58, 60, 88
solar 35
space 16, 17–18, 112, 125–6, 134
technology 35, 93
tools 32
traditional 9, 60, 112
urban 88, 89
utensils 18
Yucatecan 88
k'oben 21–2, 23, 24
Kopytoff, I. 129
Kurripako 46

Lathrap, D. W. 43
Lemonnier, P. 44
lifecycle
ceremonies 62
rituals 62
Lima 142, 143
Limón 167
food culture 171
lorena 39

MacCannell, D. 169
maicillo 30, 39 n 1
maize 19–23, 41, 63, 66 n 5, 75, 102, 155, 156
Manihot esculenta Crantz 41
manioc
bitter 42
cultivation 43
detoxification 42
domestication 43
juice 120
in Peruvian cuisine 144, 146
processing 47–9
production 182
sweet 43
toasted 113, 114
toxicity 43
mañoco 43, 45–8
Massey, D. S. 101, 161
media 112, 120, 181, 183
stars 140–1
Mérida 87, 94–5
metal mill 18, 20, 35, 36
metate 6, 7, 15, 19–21, 23, 24, 31, 32, 34, 35–6, 39, 59, 75–7, 78, 80, 105, 182
Michael, M. 4, 87
microwave oven 5, 92, 105, 117, 130, 131
milling stone (*see metate*)
Ministry of Culture of Colombia 159
modernization 4, 35, 58, 128, 162
molcajete 6, 15, 21, 59, 78, 105, 107
mole 7, 61, 63, 103, 106
moqueca 119
mortar 21, 73, 75, 91, 93
multimedia 102

INDEX

neocolonialism 29, 39
New Spain 73, 74
nixtamal (*–ized*) 20, 21, 35, 36, 78, 79
nongovernmental organizations (NGOs) 29, 35, 39, 171
nostalgia 2, 6, 7, 8, 15, 20, 99, 100, 107, 113, 182

Oaxacalifornia 101, 102
objects
    everyday 5
    material 91
    new 45
    technological 5
Oldenziel, R. 127
Orinoco River 47

pan 37, 45, 63, 75, 91, 92, 112, 132, 143
patio 17, 18, 38
Peru 139, 182
    Andean 147
Peruvian
    DNA 141
    fusion 139, 141–2, 149, 150
    gastronomic revolution 139, 140
Piaroa (Wotjuja) 46
pib 6, 18, 20, 21–2, 24, 88, 92, 93, 95
pit oven (*see pib*) 15
pot 21–2, 38, 47, 58, 63, 112, 132
    aluminum 59, 88, 92
    Capixaba 118, 119
    clay 33, 58, 118
    fondue 91
    iron 75
    terrine 93
potato 37, 99, 144
    baked 155
    boiled 144
    chips 63
    huayro 146
    mashed 63, 155
    Peruvian 148
    salad 63
    soup 156
    sweet 143, 144, 146, 174
practical knowledge 169, 173
pressure cooker 92, 95, 112, 113, 132, 133, 134
    electric (*Reina*) 132, 133

professionalization 139, 155, 156, 157
public policy 159–60
Public Policy for the Knowledge, Safeguarding and Promotion of Food and Traditional Cuisines of Colombia 159
Puerto Ayacucho 42, 45–7, 49

re-appropriation 182
refrigerator 5, 63, 92, 93, 94, 95, 112, 117, 129, 130–2, 134, 149
regional
    affiliation 9
    association 111
    belonging 161
    cookbook 86, 116, 121
    cookery style 5
    cooks 96
    cookware 119
    cuisine 55, 56, 62–4, 85, 100, 154, 156–7, 162
    culinary field 86
    dimensions of kitchens 55
    dishes 58, 120
    food 7, 87
    foodscape 88
    gastronomy 88, 93, 94
    hegemony 57
    identity 57, 86, 97, 167
    ingredients
    recipes 94
    restaurants 121
    specialties 118, 121
    taste 96
    Yucatecan cooking 85, 86
Reina (*see* pressure cooker)
religious feasts 55, 182
restaurant
    a quilo 115
    buffet-style 115, 116
    comfort food 8, 154, 155–6, 158
    ethnic 170
    fast-food 96, 104, 105, 116
    fine dining 8, 142, 147, 149, 154, 155
    fusion 116, 154
    haute cuisine 5
    history 163 n 6
    new Colombian cuisine 154
    nouveaux/nouvelle 5, 111

Peruvian cuisine 141
  in popular markets 8, 156
  upscale/high-end 148, 157
rice 41
  and beans 112, 113–16, 119, 167, 171, 173
  cooker 91, 92, 132
  as ingredient 23, 93, 133
  red 63
  and quinoa salad 155
  soup 156
ritual
  ceremonies 62
  community 61, 62
  culinary practices 181
  practices 22–3
  spaces 2, 56
  uses of food 2, 88
rondón 167, 172
  cooking class 173–7

Santiago de Cuba 129, 131, 133
sazón 78–9, 81
Schlanger, N. 4
sebucán 44, 47, 48, 50, 51
skills 49, 50, 72, 183
  cooking 60, 77, 117, 142
  corporeal 181
  culinary 140, 143
  in Istmeño cuisine 62
  knife 48
  traditional 182
skillet 92
socialism 125, 126, 128, 129, 133, 134
Spang, R. 163 n 6
stove 35, 58, 89, 92, 119
  cast-iron 75
  chef's 117
  gas 18, 22, 92
  lorena 38–9
  oven 92
  propane 112
  solar 36–8
street food vendors 6, 45, 77–9, 86, 94, 115, 158–60
sweet potato (see potato)

tamales 22, 75, 80, 92, 106, 156
  bean 63

black *mole* 61
chickpea 149
de chile 78
de dulce 78
de mareña 63
guinea pigs 149
homemade 77
mucbil pollos 20, 92
vegetarian 103
taste 181
  authentic 176
  changes 87
  for food 87, 94
  of food 1, 2, 5, 87, 99
  of home 100, 107
  homogenizing 91
  local 86
  naturalization of 96
  new 139, 140
  of Oaxaca 105
  preferences 50, 97
  regional 96
  of seven moles of Oaxaca 103
  as shared value 154
  of tortillas 104
  of Yucatecan food 7, 88, 94
  Yucatecan-Lebanese 90
techniques
  in *casabe* production 44, 48, 51
  continental 111
  cooking 4, 87, 88, 92, 95, 97, 141, 156, 168–70, 172–3, 175, 178, 183
  culinary 3, 9, 95, 150, 157
  of food production 107
  haute cuisine 141
  in *Istmeño* cuisine 56, 62
  and meaning 156–8
  modern 55, 96
  of *mole negro* 106
  new 64, 111
  preparation 100, 105, 167, 172, 176
  production 101
  traditional 55, 95, 111, 174
technological
  appropriations 6
  artifacts 44
  assemblages 87–8
  change 3, 31, 51, 86

innovation 71, 127
instruments 51
intervention 93
mediation 88
objects 5
shift 49
simplicity 31, 35, 38
system 42, 44–5, 48, 51
tools 47
technology
  ancient 73
  appropriate 29, 31
  concept of 4–5, 9, 182
  cooking 93
  culinary 15–17, 31, 35, 38, 101, 182–3
  cultural biography of kitchen 129
  definition of 4, 44, 72, 97, 157
  food 71
  intermediate 38
  kitchen 22, 35
  modern 24, 42
  Neolithic 35
  traditional 36, 42, 45
technoscape 85
technoscience 87
Tejano 75
  diet 76
  rural society 77
Tex-Mex 7, 82
thermoliths 22
tools 181
  basic 32
  in *casabe*-making 48
  cookbooks as 91
  cooking 15, 183
  culturally appropriate 5
  domestic 32
  kitchen 105
  modern 35
  new 36
  in *rondón* preparation 175
  as technology 85, 97
  traditional 59, 85, 182, 183
tortillas 22, 31–4, 63, 75–9, 182
  corn 18, 31, 35–9, 99, 105
  cultural value of 35

fiota 31, 32
galana 31, 32, 38
hand-made 20
maicillo 30
maize 20
tlayuda 102
totopos 61, 66 n 5
totopostes 40
tourist consumption 168, 169, 170, 171, 173, 177
tourist gaze 8, 169–70
translocal 3, 5, 7, 9, 85, 88, 96
Tuan, Y.-F. 86

Urry, J. 169, 170

values
  aesthetic 2
  affective 113
  of communal technologies 127
  community 155
  contemporary 96
  of convenience 94
  of cooking 116
  cosmopolitan 154
  culinary 2, 5, 55, 64, 144, 157
  cultural 35, 39, 119, 157
  emotional 31, 32
  ethical 157
  of foreign cuisines 96
  modern
  nutritional 30
  patrimonial 118
  social 32
  traditional 55, 65

Wang, N. 169
water heater 132, 133

Yanomami 7, 45, 46
Yekuana 46
yuca 43, 156
Yucatecan
  gastronomy 86
  identity 86

Zachmann, K. 127